AF541272

NANOCRYSTALS AND NANOSTRUCTURES
FROM BASIC PRINCIPLES TO APPLICATIONS

NANOCRYSTALS AND NANOSTRUCTURES
FROM BASIC PRINCIPLES TO APPLICATIONS

BY

RUBY PATEL
GAH WELLS

KNOWLEDGE BAKERS

ISBN: 9789390013241 (HB)

Published by : **Knowledge Bakers**
Pune, India
Email: knowledgebakers2019@gmail.com

Digitally Printed at : **Replika Press Pvt. Ltd.**

PREFACE

Nanomaterials and nanotechnologies have been the center of interest of recent research. Various research fields, including physics, chemists, materials science, and engineering are involved in this advanced research. A lotus plant has nanometric-sized hairs on its leaf surface that can run off water and act as self-cleaning agents. This is a hint of nanosize given by nature. The American physicist Richard Feynman in 1959, inspired in the field of nanotechnology at an American Physical Society's lecture titled "There's Plenty of Room at the Bottom." He told about the arrangement of atoms which can be put atom-by-atom as per the chemist, that help to generate a new substance, whether it is atomic scale or macro scale. Then the definition of nanotechnology has given by Norio Taniguchi, Tokyo Science University in 1970. He stated that "Nanotechnology is defined as the technology associated with different design and production of structures by controlling the size and the shape at nanometer scale." In nanotechnology, objects and phenomena are considered only at a very small scale, around 1–100 nm. Nanomaterials have distinct physical, mechanical, thermal, chemical, electronic, and magnetic properties different from those of the same chemically composed bulk materials.

Technology revolutions induce social changes and environmental apprehension. In the words of Peter Grutter, physicist, McGill University one who state their views as "I have a lot of hope that nanotechnology can get governments to put structures in place nationally and internationally that are more adaptable, so that the public could decide it wants the technology and then we could go for it" whereas Thomas Murray (a bioethicist and President at the Hastings center) states that "It is important for the earliest uses of nanotechnology to be thoughtful and attractive. The public is looking for control of their lives. Could you, for example, use nanotechnology to empower people to practice healthy behaviours?" Currently, nanotechnology is mostly used in cosmetics, medicine/drugs, fabrics, defence, energy, and water purification, but within few years all others domains will also be using nanotechnology as a potential tool.

The book provides an in-depth understanding of the basic principles of nanocrystals and nanostructures and their potential applications in various fields such as optoelectronics, catalysis, photonics, and biomedical imaging. Analyzes the most recent advances and industrial applications of different types of nanoparticles and architecture nanostructures, taking into account their unusual structures and compositions. Identifies novel nanometric particles and architectures that are of particular value for applications and the techniques required to use them effectively.

This is a valuable reference for materials scientists, chemical and mechanical engineers working both in R&D and academia who want to learn more on how nanoparticles and nanomaterials are commercially applied.

We are grateful to all those persons as well as various books, manuals, periodicals, magazines, journals etc. that helped in the preparation of this book. In spite of the best efforts, it is possible that some errors may have occurred into the compilation and editing of the book. Further queries, constructive suggestions and criticisms for the improvement of the book are always welcomed and shall be thankfully acknowledged.

Ruby Patel
Gah Wells

CONTENTS

Section 1

Synthesis of Nanocrystals

Chapter 1

Synthesis and Luminescent Properties of Silicon Nanocrystals

Abstract

Nowadays, study of silicon-based visible light-emitting devices has increased due to large-scale microelectronic integration. Since then different physical and chemical processes have been performed to convert bulk silicon (Si) into a light-emitting material. From discovery of Photoluminescence (PL) in porous Silicon by Canham, a new field of research was opened in optical properties of the Si nanocrystals (Si-NCs) embedded in a dielectric matrix, such as SRO (silicon-rich oxide) and SRN (silicon-rich nitride). In this respect, SRO films obtained by sputtering technique have proved to be an option for light-emitting capacitors (LECs). For the synthesis of SRO films, growth parameters should be considered; Si-excess, growth temperature and annealing temperature. Such parameters affect generation of radiative defects, distribution of Si-NCs and luminescent properties. In this chapter, we report synthesis, structural and luminescent properties of SRO monolayers and SRO/SiO_2 multilayers (MLs) obtained by sputtering technique modifying Si-excess, thickness and thermal treatments.

Keywords: SRO monolayers, silicon nanocrystals, sputtering, multilayers

1. Introduction

The use of photonic signals instead of electrons to transmit information through an electronic circuit is an actual challenge. Unfortunately, it is well known that bulk silicon (Si) is an indirect bandgap semiconductor, making it an inefficient light emitter. Therefore, great efforts have been taken to obtain highly luminescent Si-based materials in order to get Si-based photonic devices, especially a light-emitting device [1–3]. Such circumstances have led to explore new options for converting silicon into a luminescent material. Si-NCs embedded in a dielectric material as silicon-rich oxide (SRO) or silicon-rich nitride (SRN) show a prominent photoluminescence (PL) emission in red and blue-green region, respectively [4, 5]. Thus, SRO films have been considered as promising candidates for potential applications in Si-based optoelectronic devices, and their fully compatibility with the complementary metal-oxide-semiconductor (CMOS) processes. At present, different techniques or methods have been employed to produce SRO films or SRO/SiO_2 MLs [6–13]. In such structures, the SRO-thickness, Si-excess and annealing temperature are parameters that promote the formation of Si-NCs and radiative defects, affecting the optical and electrical properties of SRO monolayers or SRO/SiO_2 ML.

In this chapter, results about of Si-NCs embedded at SRO monolayer and SRO/SiO_2 MLs deposited by the RF Co-sputtering method as a function of Si-excess (5.2–14.3 at.%) and modulating the SRO-thickness layer (2.5–7.5 nm) of MLs are shown. Both, SRO monolayers and SRO/SiO_2 MLs have a broad emission band in the orange-red region (1.45–2.3 eV). Nevertheless, the SRO/SiO_2 MLs emit a stronger PL intensity compared with SRO monolayers. The most intense PL emission is observed in MLs when the SRO-thickness is 5 nm, and with the highest Si-excess (14.3 at.%), which is important for the design of electroluminescent devices with low threshold voltage. Although multilayer structures with good control in crystal size have been studied [14], in this chapter optical and structural properties are analyzed in detail. Furthermore a comprehensive study of synthesis of Si-NCs in SRO and SRO/SiO_2 MLs as a function of Si-excess (5.2–14.3 at.%) and annealing temperature is presented.

2. Experimental procedure

In this chapter, structural and optical properties of co-sputtered SRO monolayers and SRO/SiO_2 MLs are obtained and studied as a function of the annealing temperature and the change in the Si-excess. Both structures were deposited by the co-sputtering of Si (2″, purity of 99.999 %) and SiO_2 (2″, purity of 99.99%) targets using a Torr International magnetron sputtering system (13.56 MHz). Both structures were deposited on p-type (100) Si wafers with resistivity of 2–5 Ω-cm. Before deposition, Si substrates were cleaned in ultrasonic bath with acetone, ethanol, and deionized water successively. After being dried with nitrogen, the substrates were immediately loaded into the chamber of the sputtering system. Once a base pressure of ~1×10^{-6} Torr is achieved, Ar flow of 60 sccm is introduced into the chamber at a working pressure of 2.4 mTorr.

The SRO monolayers were deposited by the simultaneous co-sputtering of Si and fused quartz (SiO_2) targets. The Si-excess content in SRO monolayers was modified by changing the RF power applied to the Si-target (PSi) from 40 to 80 W maintaining constant the RF power applied to the SiO_2 target (P_{SiO_2} = 100 W). Thus, RF-power densities to the Si-target (PD_{Si}) were applied; 1.97, 2.47, 2.96, 3.45, and 3.94 W/cm^2 to obtain 4.4, 5.2, 10.2, 14.3, and 16.9 at.% of Si-excesses, respectively (**Table 1**). On the other hand, the RF-power density applied to the

Sample	Sample topology	Si-excess (at.%)	≈Thickness (nm)	
			As-deposited	Annealed
A	SRO monolayer	5.2	108	90
B	SRO monolayer	10.2	82	70
C	SRO monolayer	14.3	90	76

Table 1. Thickness and Si-excess obtained for SRO monolayers.

Sample	Sample topology	Si-excess (at.%)	Total ≈ Thickness (nm)	
			As-deposited	Annealed
1A	~2.5 nm-SRO/~6 nm-SiO_2		107	105
2A	~5 nm-SRO/~6 nm-SiO_2	**5.2**	135	127
3A	~7.5 nm-SRO/~6 nm-SiO_2		165	150
1B	~2.5 nm-SRO/~6 nm-SiO_2		99	103
2B	~5 nm-SRO/~6 nm-SiO_2	**10.2**	130	120
3B	~7.5 nm-SRO/~6 nm-SiO_2		170	152
1C	~2.5 nm-SRO/~6 nm-SiO_2		107	102
2C	~5 nm-SRO/~6 nm-SiO_2	**14.3**	140	124
3C	~7.5 nm-SRO/~6 nm-SiO_2		168	154

Table 2. SRO and SiO_2 thickness and Si-excess in the SRO/SiO_2 multilayer structures.

SiO_2 target (PD_{SiO_2}) was 5.09 W/cm^2. For SRO/SiO_2 MLs, SRO interlayers with the same Si-excess that SRO monolayers were used. First, a SiO_2 layer was deposited onto the silicon substrate followed by a SRO film to obtain a SRO/SiO_2 bi-layer. Ten periods of SRO/SiO_2 bi-layers were deposited with an additional (upper) SiO_2 film (10 nm) to avoid oxidization during high temperature annealing. Each SiO_2 layer was about 6 nm in thick while the SRO layer thickness was modified from 2.5 to 7.5 nm (see **Table 2**).

All samples were deposited at 100°C. However, Si-NCs in SRO monolayers and SRO/SiO_2 MLs were produce by thermal annealing in a conventional tube furnace between 900 and 1200°C in N_2 environment for 2 h. Thickness of the films was measured by reflectance using a Filmetrics F20UV equipment. The chemical compositions and Si-excess of samples were analyzed by a Thermo Scientific X-ray photoelectron spectroscopy (XPS) Escalab 250Xi equipment. The Si oxide phase was studied by Fourier transform Infra-red (FTIR) spectroscopy using a Bruker Vector 22 spectrometer in the range 400–4000 cm^{-1}. The PL emission spectra were measured with a Horiba Fluoromax 3 system. The samples were excited using a 300 (4.13 eV) nm radiation and the PL emission signal was collected from 400 to 900 nm (1.37–3.1 eV) with a resolution of 1 nm.

3. Deposition rate Si and SiO_2 films

The first step to synthesize silicon nanocrystals embedded in SRO films by Co-sputtering is to consider the deposition rate of Si and SiO_2 films. Such procedure allows to obtain RF powers appropriate to Si and SiO_2 targets and silicon content suitable for SRO films. Some experimental studies report RF powers applied to the Si higher than SiO_2 target [15, 16]. However, it is known that the deposition rate of SiO_2 is low (0.1Å/sec) and high silicon content could be present in SRO films if the deposition rate of Si is much higher than the SiO_2 film. This effect is shown in **Figure 1**; where Si and SiO_2 films were individually deposited for 1200 s using Si and SiO_2 targets, respectively.

Figure 1 shows the deposition rate and thickness of Si and SiO_2 films as function of the RF-power density (PD) applied to Si and SiO_2 targets. It can be observed that the deposition rate for SiO_2 is lower (0.15 Å/s) at 100 W while that Si films (0.12 Å/s) at 50 W, where similar thickness (~16–18 nm) is deposited with a PD_{Si} = 2.47 W/cm² for a Si film and PD_{SiO_2} = 5.09 W/cm² for a SiO_2 film. Therefore, to ensure Si-excess similar to the reported by other techniques [17–21], SRO films were deposited considering RF-power densities between 1.97 and 3.94 W/cm² applied to the Si-target and a PD_{SiO_2} constant of 5.09 W/cm². Selected the appropriate power densities, SRO monolayers and SRO/SiO_2 ML were deposited at 100°C and annealed at 1100°C.

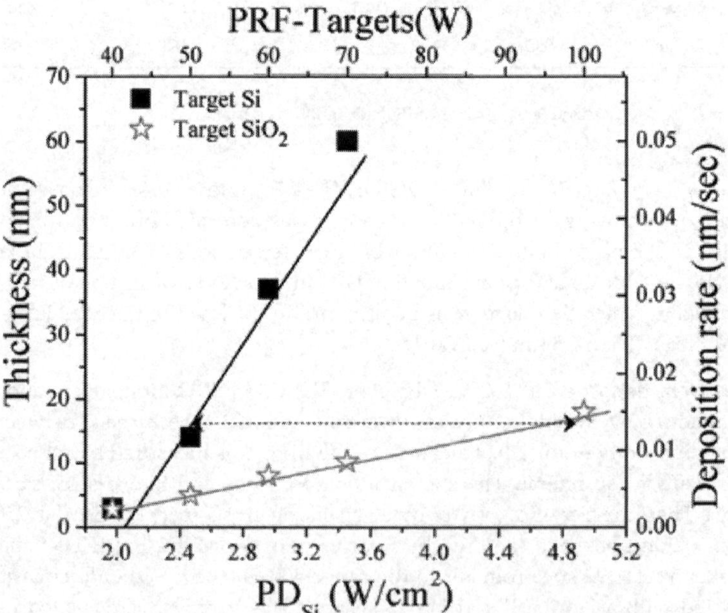

Figure 1. Deposition rate and thickness of Si and SiO_2 films as function of the RF-power density. From Coyopol et al. [25].

4. Atomic composition and XPS-Si2p of SRO monolayers and SRO/SiO_2 multilayers

4.1. SRO monolayers

The optimization of sub-stoichiometric silicon-rich oxide (SRO) from precise control of Si content in SRO films is an important parameter to tune the electrical and optical properties of luminescent and photovoltaic devices. In this regard, it has been reported that there might be an optimal Si-excess to find an efficient photoluminescence (PL) emission [18, 19]. The PL quantum yield of SRO films ($1.4 < x < 1.9$) annealed at 1100°C, increase significantly for a Si content x = 1.8 (2.5 at.% of Si-excess) [18]. For this Si-excess, Si-NCs are not detected by HRTEM, however intense emission in red region was observed. It has been suggested that for a medium Si-excess (<9 at.%), oxygen-related defects, such as Si=O (nonbridged

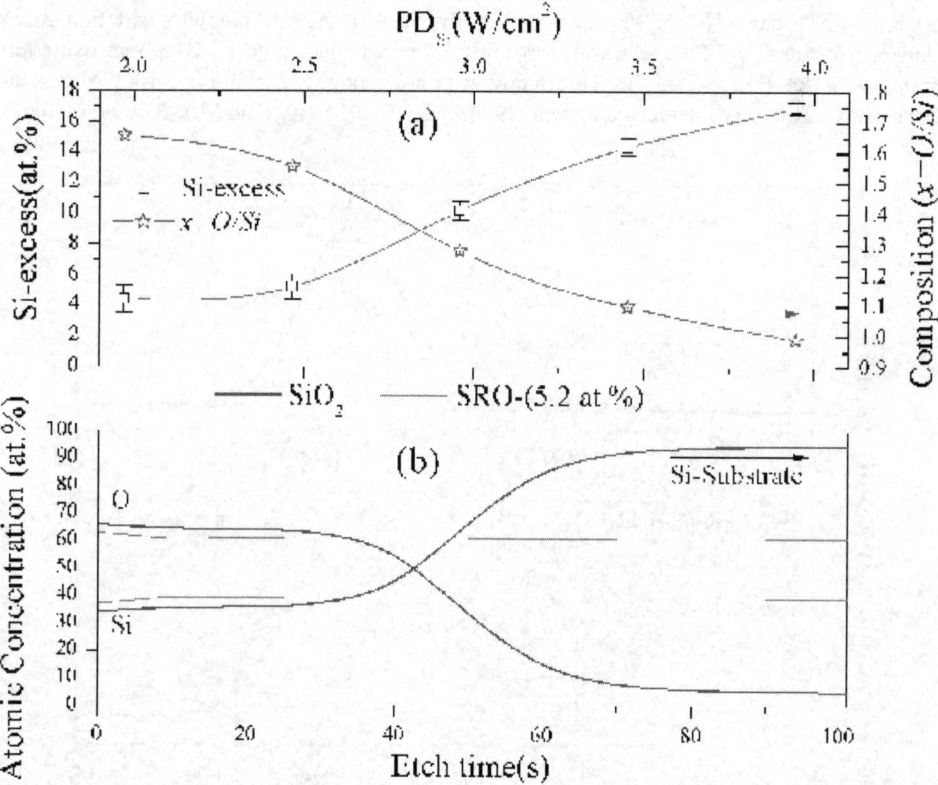

Figure 2. (a) Si-excess and x in SRO as a function of the power density applied to Si-target. (b) XPS in-depth profile of SRO monolayer with 5.2 at.% of Si-excess respect to SiO_2 film.

oxygen passivation) and/or ultra-small oxidize Si-NCs (diameters <2 nm) are the most optically actives for promote an intense emission in red region (1.5 eV). Thus, a considerable Si-excess (>10 at.%) in SRO films is not necessary to obtain a maximum PL emission.

In this chapter Si-excess between 4.4 and 16.9 at.% are reported. In order to know the composition (x = O/Si) and Si-excess of SRO monolayers and SRO/SiO_2 multilayers, XPS measurements in-depth profile were performed. In **Figure 2b**, the analysis in-depth profile of an SRO film with 5.2 at.% of Si-excess compared to a SiO_2 film is shown. It can be seen that silicon content in SRO film increase respect to SiO_2 film. The atomic concentration of Si content in SRO film was ≈38.5 at.% respect to the SiO_2 film with an atomic concentration of ≈33.33 at.%. Such effect demonstrates Si-excess in SRO monolayers. For all SRO monolayers Si-excess increases from 4.4 to 16.9 at.% as the power density of Si-target rise from 1.97 to 3.94 W/cm^2, respectively (see **Figure 2a**). Where, Si-excess of 4.4, 5.2, 10.2, 14.3, and 16.9 at.% were found for PD_{Si} values of 1.97, 2.47, 2.96, 3.45 and 3.94 W/cm^2, respectively.

The presence of Si-excess in SRO films is also detected by the oxidation states of Si obtained from Si2p-XPS data. This fact is suggested on the basis of the intermediate oxidation states detected in the Si2p-XPS spectra. According to Si-excess measured by XPS, five oxidation states were detected in samples, before and after annealing. For example for a SRO monolayer with 5.2 at.%, peaks located around 99.6 eV, 100.5, 101.5, 102.5 and 103.5 eV were found

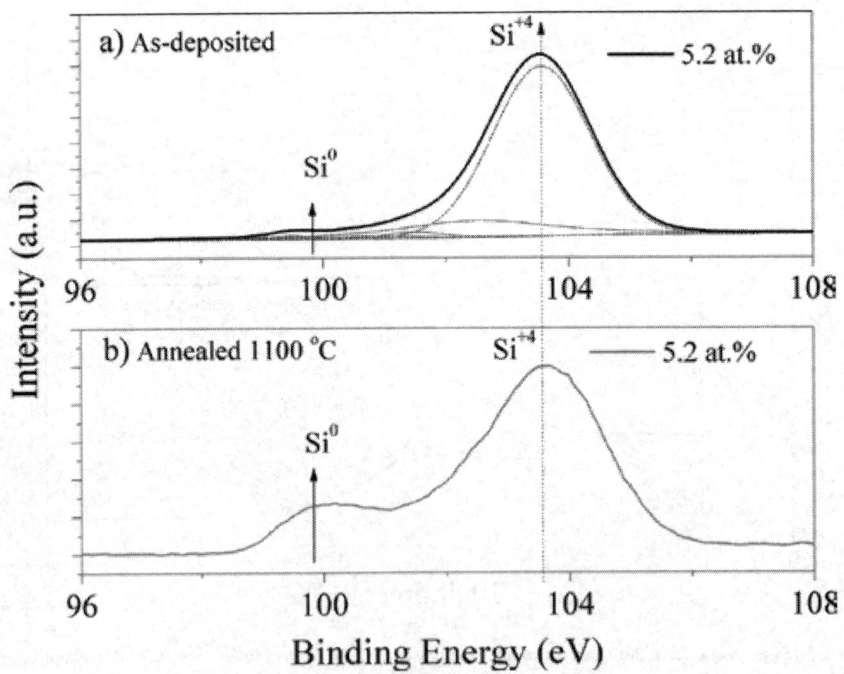

Figure 3. Si2p-XPS peaks of SRO monolayer (5.2 at.% of Si-excess) (a) as-deposited and (b) annealed at 1100°C.

(**Figure 3a**). The aforementioned peaks correspond to the Si^0 (elemental Si), Si^{+1} (Si_2O), Si^{+2} (SiO), Si^{+3} (Si_2O_3) and Si^{+4} (SiO_2) oxidation states, respectively [22, 23]. The contributions of the intermediate Si peaks (Si^{+1}, Si^{+2}, Si^{+3}) corresponding to the suboxide phases become smaller when the SRO films are thermally annealed at 1100°C, while Si^{+4} peak for SiO_2 and Si^0 peak for elemental Si becomes dominant, especially the intensity of Si^0 peak increases (**Figure 3b**), explaining the Si-SiO_2 phase separation and Si-NCs formation after annealing, as shown by similar results found by Chen et al. [24].

4.2. SRO/SiO_2 MLs

For the design of SRO/SiO_2 ML structures, SRO layers with the same Si-excess that monolayers were used. In **Figure 4a**, the composition in-depth profile for an as-deposited SRO/SiO_2 MLs with 5 nm-thick SRO layer corresponding to 5.2 at.% and 10.2 at.% of Si-excess is shown. As we can see, different changes in atomic concentration (x) are observed corresponding to different SRO-SiO_2 interfaces as the etch time increases. A better contrast between SRO and SiO_2 stoichiometry is observed when the Si-excess and SRO-thickness is increased to 14.3 at.% and 7.5 nm, respectively (sample 3C, before and after annealing), as observed in **Figure 4b**. A densification in thickness is observed in this multilayer after thermal annealing and it is attributed to a microstructural reordering in SRO layers.

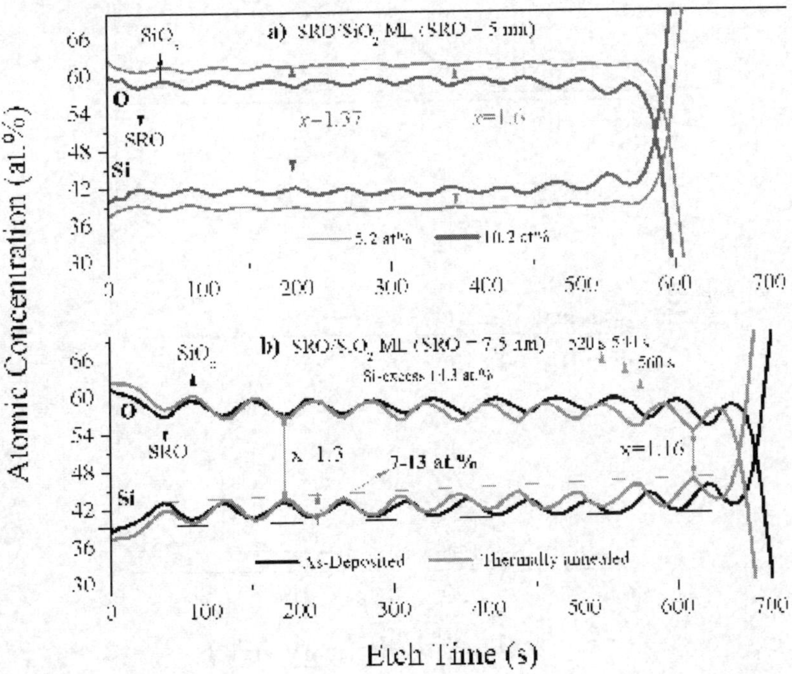

Figure 4. XPS in-depth profile of (a) as-deposited multilayer (SRO-5 nm) with 5.2 at.%, 10.2 at.% of Si-excess and (b) multilayer (SRO-7.5 nm) with 14.3 at.%-before and after thermal annealing (1100°C). From Coyopol et al. [28].

The atomic concentrations (x) obtained in multilayers as-deposited ($x = 1.37$, $x = 1.6$) are near to SRO monolayers ($x = 1.3$, $x = 1.5$). Such difference or reduction of atomic concentration could be caused by the etch time performed by XPS. A reduction of the atomic concentration (x) in sample 3C by effect annealing is observed (**Figure 4b**), even this effect being more noticeable near the substrate ($x = 1.3 \rightarrow x = 1.16$). In this way, it is likely that a re-diffusion of Si atoms between SRO-SiO_x interfaces as well as within SRO layers after thermal treatment is performed. Such annealing, promotes crystallization of Si clusters and Si-NCs formation between SRO-SiO_x interfaces [25–27]. The Si-excess or Si clusters within of multiple SRO layers were detected from Si2p-XPS analysis. For this study, sample 3C was analyzed specifically in the etch time range of 520–560 s (see **Figure 4b**), specifically in critical points; SiOx layer (520 s), near the SRO-SiOx interfaces (544 s) and SRO layer (560 s) as is shown in **Figure 5**.

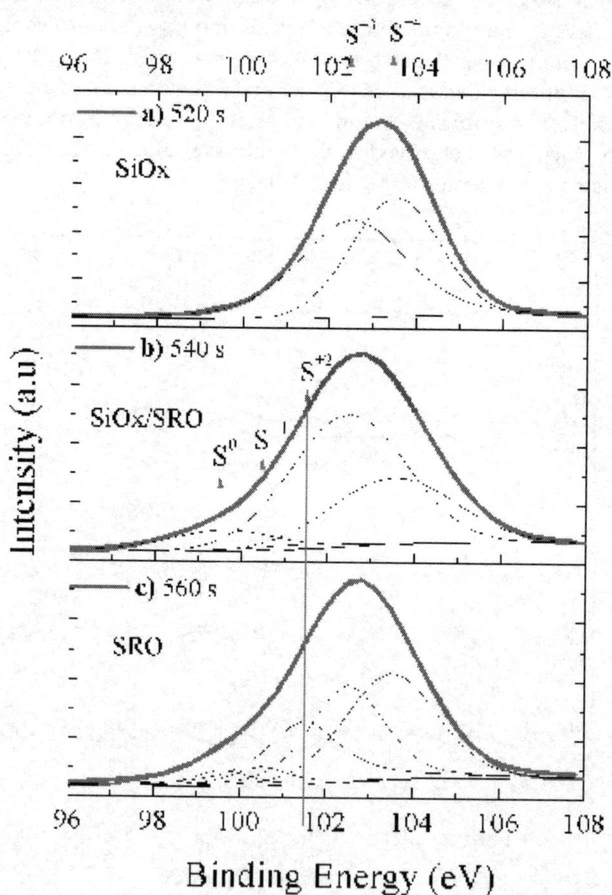

Figure 5. Si2p-XPS peaks of multilayer (SRO-7.5 nm) with 14.3 at.%- thermal annealing (1100°C) with a etch time range analysis at (a) 520 s (SiO_x layer), (b) 544 s (SRO-SiO_x interfaces) and (c) 560 s (SRO layer).

It can be observed that for etch time at 560 (SRO) and 540 s (SRO-SiO_x), five oxidation states were detected, located around 99.6 eV, 100.5, 101.5, 102.5 and 103.5 eV (**Figure 5b** and **c**). For Si2p-XPS analysis in 520 s (SiO_x), the oxidation state Si^0 was not detected (**Figure 5a**), only oxidation states Si^{+3} and Si^{+4} were detected, so that Si atoms diffusion is unlikely in this region. Thereby, for an etch time at 544 and 560 s, five oxidation states were perfectly adjusted, even the Si^0 state is greater in 544 s, which could be tentatively edges SRO-SiO_x interface. Therefore Si-diffusion is likely even in the SRO-SiO_x interfaces and SRO nanolayers. This study evidences crystallization of silicon clusters after annealing and formation of Si-NCs in the SRO-SiO_2 interfaces and the SRO nanolayers by Si atoms diffusion. This effect was demonstrated before by HRTEM measurements [25, 28].

5. FTIR and effect annealing of SRO monolayers and SRO/SiO_2 MLS

5.1. SRO monolayer

The nature of the chemical bonds and the phase separation of Si and SiO_2 of SRO monolayers before and after thermal annealing are studied by FTIR. **Figure 6a**, shows FTIR spectrum of as-deposited SRO monolayers with 5.2, 10.2 and 14.3 at.% of Si-excess. As shown in **Figure 6a**, three characteristic rocking, bending, and asymmetric stretching Si-O-Si vibrational modes of SiO_2 are

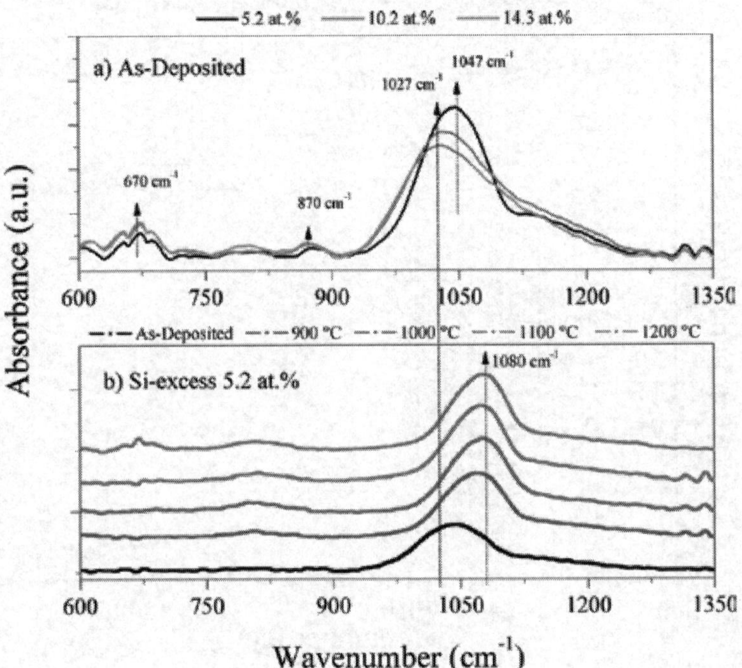

Figure 6. FTIR spectra of SRO monolayers (a) As-deposited and (b) Thermally annealed at 900°C, 1000°C, 1100°C, 1200°C.

observed at around 446–450, 800–872, and 1026–1040 cm^{-1}, respectively [29]. The Si-O-Si stretching band shifts toward low wavenumbers (1047–1027 cm^{-1}) as the silicon content increases (from 5.2 to 14.3 at.%) in as-deposited monolayers. An additional IR band of low intensity is observed around 880 cm^{-1}. Such vibrational mode disappears when samples are thermally annealed, indicating a rearrangement of the SiO_x network due to a diffusion of Si atoms, which promotes Si-SiO_x phase separation and the formation of Si-NCs in SRO layers [25–27]. **Figure 6b**, shows the effect of annealing between 900 and 1100°C on SRO monolayer of 5.2 at.% of Si-excess, where the disappearance of band located around 880 cm^{-1} is observed. Likewise, a phase separation and Si-NCs formation in SRO monolayers is corroborated by the shift observed in the Si-O-Si stretching band position (1047–1080 cm^{-1}) after thermal annealing [25, 27]. It can be seen that for an annealing above about 1000°C, stretching peak appears at 1080 cm^{-1}. Such effect means that temperatures above 1000°C are sufficient for Si-SiO_x phase separation and Si-NCs formation in the SRO monolayers. It is worth mentioning that an annealing temperature of 1100°C by 2 h provided better results in the structural and optical properties [25, 28].

5.2. SRO/SiO_2 multilayers

Figure 7 shows the evolution of the FTIR spectra in the range of 900–1280 cm^{-1} for SRO/SiO_2 MLs (14.3 at.% of Si-excess) with a SRO-thickness of 2.5 (sample 1C), 5 (sample 2C), and

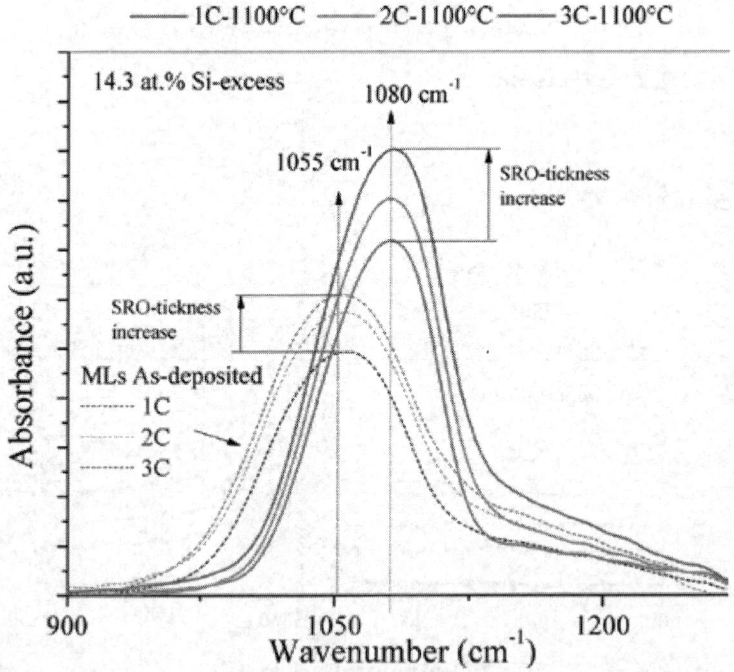

Figure 7. FTIR spectra of SRO/SiO_2 MLs as function of SRO-thickness (2.5–7.5 nm) of as-deposited and thermally annealed at 1100°C.

7.5 nm (sample 3C), before and after thermal annealing (1100°C). For MLs as-deposited, the stretching peak position remains at 1055 cm^{-1} due to the same Si-excess, 14.3 at.%. However, intensity of the stretching peak is modified as thickness of SRO intermediate layers increase (2.5–7.5 nm), which is to be expected since the thickness of the MLs increases (**Table 2**) and the number Si-O-Si bonds of these molecules increases. The shift in the Si-O-Si stretching band was also observed from 1055 to 1080 cm^{-1} after the thermal annealing. This effect, again, indicates a phase separation (Si-SiO_x) and therefore the Si-NCs formation within the SRO-bulk layers and possibly in the SRO-SiO_2 interfaces. By other hand, all SRO/SiO_2 MLs exhibited narrower Si-O-Si stretching bands respect to the SRO monolayers. Such effect at SRO/SiO_2 MLs can be related to a higher structural order in the SRO matrix due to a compressive stress in Si-NCs [30]. In MLs structures, the compressive stress exerted on the Si-NCs is higher unlike monolayers, where the structural order of the SRO matrix is the lowest resulting in a broad IR spectrum. Thus, it is likely that smaller nanocrystals (diameters <3 nm) must be confined in the SRO/SiO_2 MLs due to the high stress.

6. PL of SRO monolayers and SRO/SiO_2 MLs

6.1. SRO Monolayers

Figure 8a shows the PL spectra of SRO monolayers thermally annealed at 1100°C for different Si-excess. It can be seen that when Si-excess increases of 4.4–16.9 at.%, the PL emission intensity is modified. The emission intensity of this band (red region) increases considerably for the SRO monolayer with 5.2 at.% thermally annealed at 1100°C. It has been reported that SRO films deposited by LPCVD or MBE with medium Si-excess (~2.5–5 at.%) produces an efficient PL emission, however Si-NCs have not been observed [17–21]. In our case, Si-NCs were observed from a Si-excess of 5.2 at.% (x = 1.54), as it was shown in images HRTEM [25]. The PL emission in SRO films with 14.3 and 16.9 at.% is still perceptible, however is weak due to the high silicon content, even close to SiO film (x = 1). In this case, PL intensity decreases due to an increase of Si-NCs size, larger to the Bohr radius or to annihilation of radiative centers. Regarding to the PL emission observed in sample whit 4.4 of Si-excess (PD_{Si} = 1.97). A moderate emission is observed in the **Figure 8a**. According to the low concentration in this sample (x = 1.64), such emission is attributed to the formation of ultra-small oxidized Si-NCs as well as some point defects [18, 19]. It has been suggested that for a medium Si-excess (<9 at.%), oxygen-related defects, such as Si=O (Nonbridged oxygen passivation) and/or ultra-small oxidized Si-NCs (diameters <2 nm) are the most optically active centers that promote an intense emission in red region (~1.5–1.7 eV). Accordingly, Si-NCs with diameters larger than 3 nm (Si-excess >10 at.%) are not direct light-emitting centers [18, 19]. Thus, a considerable Si-excess (>10 at.%) in SRO films is not necessary to obtain a maximum PL emission.

The effect of the annealing temperature (900–1200°C) on the PL properties of the SRO monolayers also was analyzed (**Figure 8b**). The best PL emission intensity was obtained for the sample with Si-excess of 5.2 at.% (x = 1.54) and annealed at 1100°C. Thus, for the other annealing temperatures, PL intensity decreases, so a temperature of 1100° C is ideal for obtaining a maximum PL emission in SRO monolayers obtained by RF Co-Sputtering.

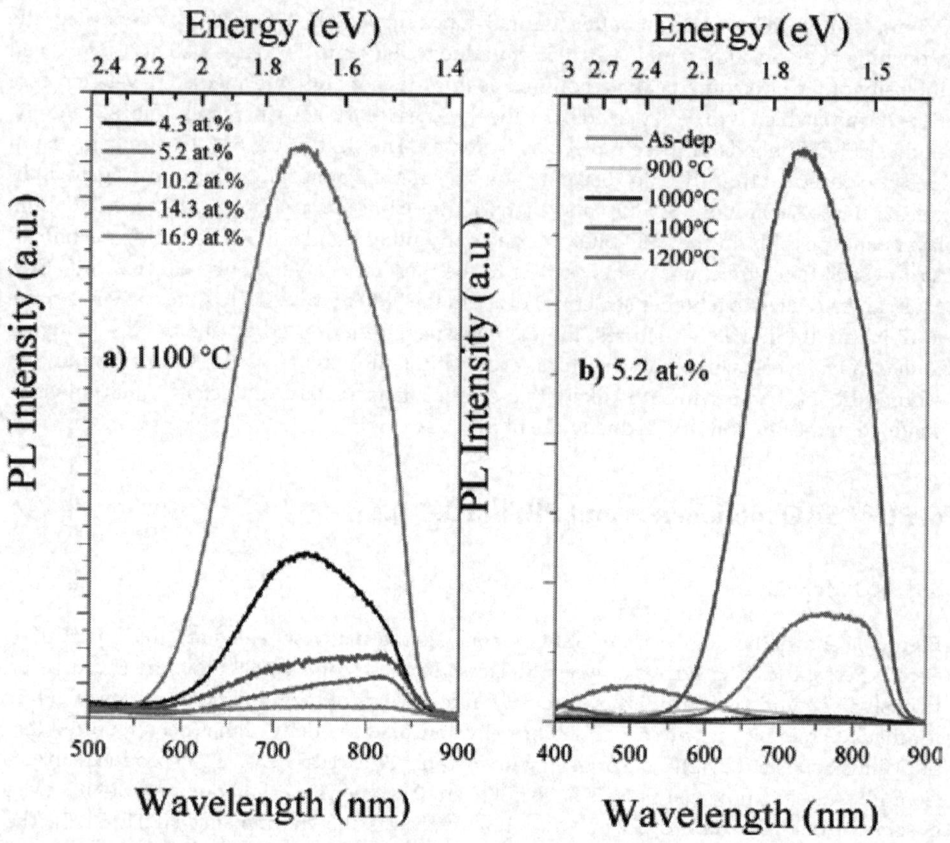

Figure 8. PL spectra of SRO monolayers (a) thermally annealed at 1100°C for different Si-excess and (b) annealed at different temperatures with 5.2 at.% of Si-excess.

6.2. SRO/SiO_2 MLs

The maximum PL emission in the SRO monolayers is obtained with a Si-excess between 5.2 and 14.3 at.% and heat treated at 1100° C. In this way for the design of the structures MLs, Si-excess of the same order was used. Both, monolayers and multilayers, emit an intense and broad emission band in the red-orange region (1.45–2.1 eV). **Figure 9b** and **a**, shows, respectively, the PL spectra of thermally annealed SRO monolayers and SRO/SiO_2 MLs with a SRO-thickness of 5 nm with different Si-excesses [28]. In this case it was found that the MLs with a thickness of 5 nm in the SRO layer, has a maximum of PL emission independent of the Si-excess. For this, an analysis of PL intensities was made considering the ratio of maximum peaks of intensities of the SRO/SiO_2 MLs and SRO monolayer. Such analysis to determine the MLs with greater emission as function of Si-excess and SRO-thickness layer. The ratio of emission intensities was calculated taking the maximum PL intensity of each peak (~1.72, ~1.6 and ~1.52 eV) that form the PL spectrum of each SRO/SiO_2 ML (I_{PL} ML) divided by the maximum PL intensity of SRO monolayers (I_{PL} Mono) located at the same energy [28]. So, the

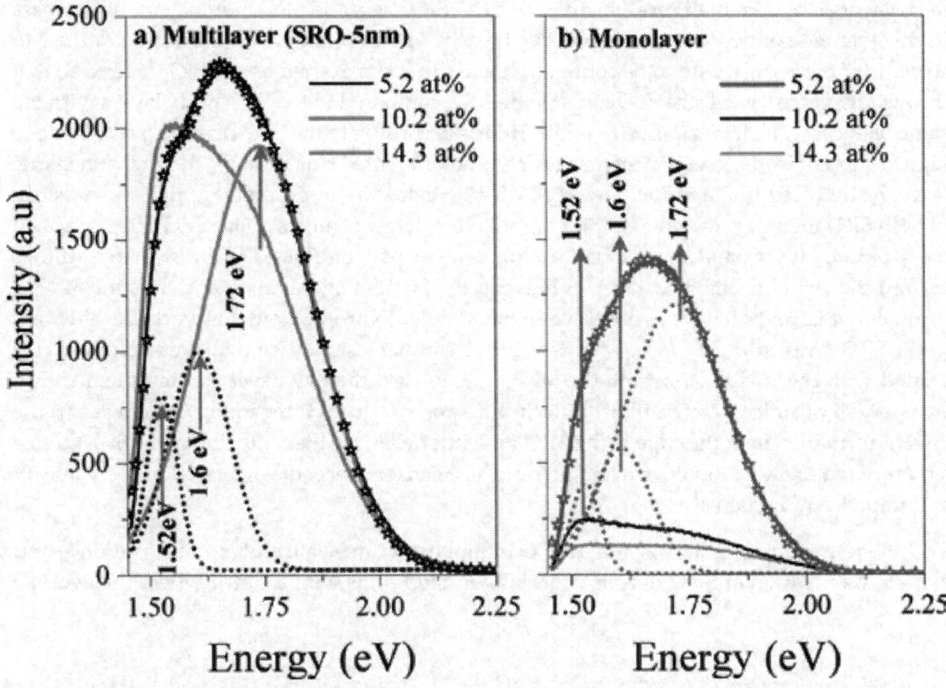

Figure 9. PL spectra of (b) SRO/SiO_2 MLs (SRO-5 nm) with 5.2, 10.2, and 14.3 at.% of Si-excess thermally annealed at 1100°C and (a) SRO monolayers. From Coyopol et al. [28].

most intense photoluminescence is achieved using SRO layers with 5 nm thick and 14 at.% of Si-excess. Such effect is important for the design of electroluminescent devices, particularly for supplying low voltages, since a large number of Si-NCs are generated as Si-excess increases.

The effect of confining Si-NCs in the SRO/SiO_2 MLs is to obtain nanocrystals sizes <3 nm due to the high stress, result in Si=O defects formation. Regarding the presence of Si-NCs and their PL properties, Wolkin et al. suggested that light emission in red-orange region originates from Si=O bonds, where oxidation can stabilize the energy position of the PL peak (~1.5–1.7 eV) from small Si-NCs (<3 nm) [31]. The model of Si=O bonds has been also used to explain the PL emission in connection to Si-NCs in a SiO_2 matrix [32]. It was suggested later that the light-emitting centers can be stabilized not only at Si-NCs surface but also in a disordered network [33], which could be promoted from the phase separation of Si and SiO_2. The emission of the Si=O defect is around 1.5–2.1 eV and it strongly affect the PL intensity when the nanocrystal size is below 3 nm [31].

7. Si-NCs in SRO monolayer and SRO/SiO_2 MLs

The formation of Si-NCs in SRO monolayers and SRO/SiO_2 multilayers was demonstrated by HRTEM analysis [25, 28]. A reduction in nanocrystal size in SRO/SiO_2 MLs respect to SRO monolayers is shown in **Figure 10a**. In both, SRO/SiO_2 MLs and SRO monolayers, an increase in the average nanocrystal size is observed as the Si-excess increases. However, in the ML structures, the Si-NCs tend to be confined due to the stress exerted by the SiO_x matrix, so that the average nanocrystal size decreases in the MLs compared to the SRO monolayers with the same Si-excess. The increase in size of the Si-NCs according to the Si-excess is expected due to high Si content leads an enhanced in the agglomeration of Si clusters after thermal annealing. It can be deduced that a reduction in Si-NCs size below 3 nm promotes a high PL emission in both SRO monolayers and SRO/SiO_2 multilayers. Recent studies about SRO films confined by two SiO_2 layers have shown that Si atoms in excess from the SRO films mainly diffuse toward the center of the SRO layer, enhancing the Si atom aggregation and thus the Si-NCs formation [8]. Nevertheless, as observed in the XPS depth profile from this work, the Si-excess in the SRO films from SRO/SiO_2 MLs reduces as compared to SRO monolayers. This effect is related with the Si-diffusion from the SRO films toward the SiO_2 layer making them silicon rich instead of stoichiometric films. This Si-diffusion produces three important effects: (i) the Si-NCs formation near the edge of the SRO-SiO_2 interfaces, (ii) the reduction of the Si-NCs size as compared to SRO monolayers and (iii) and as a consequence, the increased Si-NCs density and improved PL intensity.

The phenomenon of light emission in SRO monolayers can be attributed to the presence and complete activation of Si=O defects especially in SRO films with 5.2 at.% where the average

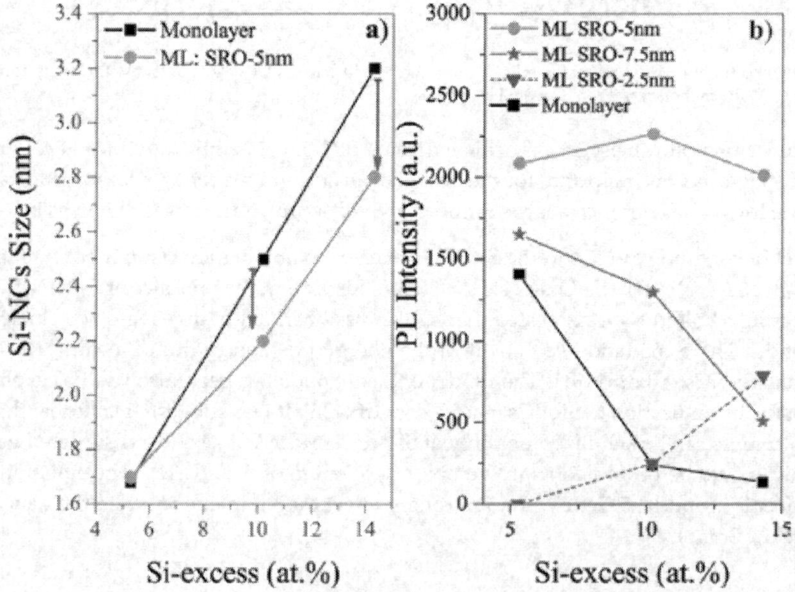

Figure 10. (a) Si nanocrystal size and (b) PL intensity in SRO monolayers and SRO/SiO_2 MLs as a function of Si-excess. From Coyopol et al. [28].

Si-NC size is about 1.68 ± 0.2 nm, as reported in other work [25]. In SRO/SiO_2 MLs, Si-NCs with diameters below 3 nm are necessary to obtain the activation of the Si=O luminescence centers. Therefore, SRO/SiO_2 MLs with SRO-5 nm where Si-NC sizes of about 1.7 ± 0.1, 2.1 ± 0.08 and 2.8 ± 0.09 nm were obtained with Si-excess of 5.2, 10.2 and 14.9 at.%, respectively [28]. The PL intensity for multilayers with SRO-thickness of 5 nm remains almost at the same order as the Si-excess increases, as shown in **Figure 10b**. However, the PL intensity of SRO/SiO_2 MLs with SRO-2.5 nm and SRO-7.5 nm and SRO monolayers decreases as the Si-excess increases. It was observed that the Si-NCs density in SRO/SiO_2 MLs is higher than SRO monolayers produced by a better control of the NCs formation in SRO nanolayers confined between SiO_2 layers. Moreover, Si-NCs with sizes below 3 nm and a large number of Si=O defects are created in SRO/SiO_2 MLs with SRO-thickness of 5 nm. These effects explain the enhanced PL intensity of multilayers making them suitable for the development of silicon-based light sources.

8. Conclusion

In the synthesis of Si-NCs by RF Co-sputtering in SRO monolayers or SRO/SiO_2 multilayers, it is important to consider parameters such as Si-excess, annealing temperature and SRO-thickness layer for SRO/SiO_2 MLs design. These parameters are important for the obtaining of a high emission either in SRO monolayers or SRO/SiO_2 MLs. It can be deduced that a high emission in both; SRO monolayers and SRO MLs is obtained for a medium Si-excess (5.2 at%). However, an important parameter to consider is the Si nanocrystal size. That is, while synthesis of Si-NCs is below 3 nm, a complete activation of Si=O defects are guaranteed. In this way it is possible to obtain a high emission in multilayers with a high Si content. For MLs where the intermediate SRO layer is 5 nm and Si-excess of 14.3 at%, a high emission is obtained due to the confinement of nanocrystals less than 3 nm. Such effect is important for the design of electroluminescent devices, particularly for supplying low voltages, since a large number of Si-NCs are generated as Si-excess increases.

Acknowledgements

Authors want to thank Oscar Solís, Cesar Leyva and Luis Gerardo Silva from CIMAV for the FIB preparation, TEM and XPS measurements, respectively. A. Morales Sánchez acknowledges the support received from CONACYT-CB #180992 and CONACYT-Scientific Infrastructure #269359.

Conflict of interest

There is no conflict of interest with the other authors.

References

[1] Tseng C-K, Lee M-CM, Hung H-W, Huang J-R, Lee K-Y, Shieh J-M, Lin G-R. Silicon-nanocrystal resonant-cavity light emitting devices for color tailoring. Journal of Applied Physics. 2012;**111**(7):074512. DOI: 10.1063/1.3702793

[2] Zhiping Z, Bing Y, Jurgen M. On-chip light sources for silicon photonics. Light: Science & Applications. 2015;**4**:e358. DOI: 10.1038/lsa.2015.131

[3] Berencén Y, Illera S, Rebohle L, Ramírez JM, Wutzler R, Cirera A, Hiller D, Rodríguez JA, Skorupa W, Garrido B. Luminescence mechanism for Er^{3+} ions in a silicon-rich nitride host under electrical pumping. Journal of Physics D: Applied Physics. 2016;**49**(8):085106. DOI: 10.1088/0022-3727/49/8/085106

[4] Shih C-F, Hsiao C-Y, Su K-W. Enhanced white photoluminescence in silicon-rich oxide/SiO_2 superlattices by low-energy ion-beam treatment. Optics Express. 2013;**21**(13):15888-15895. DOI: 10.1364/OE.21.015888

[5] Vlasukova LA, Komarov FF, Parkhomenko IN, Milchanin OV, Makhavikou MA, Mudryi AV, Żuk J, Kopychiński P, Togambayeva AK. Visible photoluminescence of non-stoichiometric silicon nitride films: The effect of annealing temperature and atmosphere. Journal of Applied Spectroscopy. 2015;**82**(3):386-389. DOI: 10.1007/s10812-015-0117-9

[6] Han PG, Ma ZY, Wang ZB, Zhang X. Photoluminescence from intermediate phase silicon structure and nanocrystalline silicon in plasma enhanced chemical vapor deposition grown Si/$SiO_{(2)}$ multilayers. Nanotechnology. 2008;**19**(32):325708. DOI: 10.1088/0957-4484/19/32/325708

[7] Kim KJ, Moon DW, Hong S-H, Choi S-H, Yang M-S, Jhe J-H, Shin JH. In situ characterization of stoichiometry for the buried SiOx layers in SiOx/SiO_2 superlattices and the effect on the photoluminescence property. Thin Solid Films. 2005;**478**:21-24

[8] Yoon JH. Enhanced light emission from Si nanocrystals produced using SiOx/SiO_2 multilayered silicon-rich oxides. Applied Surface Science. 2015;**344**:213-216

[9] Lin YH, Wu CL, Pai YH, Lin GR. A 533-nm self-luminescent Si-rich SiNx/SiOx distributed Bragg reflector. Optics Express. 2011;**19**(7):6563-6570

[10] Wu CL, Lin YH, Lin GR. Narrow-Line width and Wavelength-Tunable Red-Light Emission From an Si-Quantum-Dot Embedded Oxynitride Distributed Bragg Reflector. IEEE Journal of Selected Topics in Quantum Electronics. 2012;**18**(6):1643-1649

[11] Alarcon-Salazar J, Zaldívar-Huerta IE, Aceves-Mijares M. Electrical and electroluminescent characterization of nanometric multilayers of SiOX/SiOY obtained by LPCVD including non-normal emission. Journal of Applied Physics. 2016;**119**:215101

[12] Kansawa Y, Hayashi S, Yamamoto K. Raman spectroscopy of Si-rich SiO_2 films: possibility of Si cluster formation. Journal of Physics. Condensed Matter. 1996;**8**:4823

[13] Tomozeiu N, van Hapert JJ, van Faassen EE, Arnoldbik W, Vredenberg AM, Habraken FHPM. Structural properties of a-SiOx layers deposited by reactive sputtering technique. Journal of Optoelectronics and Advanced Materials. 2002;**4**:513

[14] Zacharias M, Yi LX, Heitmann J, Scholz R, Reiche M, Gosele U. Size-controlled Si nanocrystals for photonic and electronic applications. Solid State Phenomena. 2003; **94**:95-104

[15] Kenji I, Masahiko I, Minoru F, Shinji H. Nonlinear optical properties of Si nanocrystals embedded in SiO_2 prepared by a cosputtering method. Journal of Applied Physics. 2009;**105**:093531

[16] Hao XJ, Cho EC, Flynn C, Shen YS, Park SC, Conibeer G, Green MA. Synthesis and characterization of boron-doped Si quantum dots for all-Si quantum dot ándem solar cells. Solar Energy Materials & Solar Cells. 2009;**93**:273-279

[17] Aceves-Mijares M, González-Fernández AA, López-Estopier R, Luna-López JA, Berman-Mendoza D, Morales A, Falcony C, Domínguez C, Murphy-Arteaga R. On the origin of light emission in silicon rich oxide obtained by low-pressure chemical vapor deposition. Journal of Nanomaterials. 2012;**2012**:890701. DOI: 10.1155/2012/890701

[18] Nikitin T, Velagapudi R, Sainio J, Lahtinen J, Räsänen M, Novikov S, Khriachtchev L. Optical and structural properties of SiOx films grown by molecular beam deposition: Effect of the Si concentration and annealing temperature. Journal of Applied Physics. 2012;**112**:094316

[19] Khriachtchev L, Ossicini S, Iacona F, Gourbilleau F. Silicon nanoscale materials: From theoretical simulations to photonic applications. International Journal of Photoenergy. 2012;**2012**:872576. DOI: 10.1155/2012/872576

[20] Morales A, Domínguez C, Barreto J, Riera M, Aceves M, Luna JA, Yu Z, Kiebach R. Spectroscopical analysis of luminescent silicon rich oxide films. Revista Mexicana de Física. 2007;**S53**:279

[21] Dong D, Irene EA, Young DR. Preparation and some properties of chemically vapor-deposited Si-rich SiO_2 and Si_3N_4 Films. Journal of the Electrochemical Society. 1978;**125**:819-823

[22] López-Estopier R, Aceves-Mijares M, Falcony C. Cathodo- and photo-luminescence of silicon rich oxide films obtained by LPCVD. In: Yamamoto N, editor. Cathodoluminescence. Rijeka, Croatia: InTech; 2012. p. 324

[23] Zhang WL, Zhang S, Yang M, Liu Z, Cen ZH, Chen T, Liu D. Electroluminescence of as-sputtered silicon-rich SiOx films. Vaccum. 2010;**84**:1043-1048

[24] Chen XY, Lu YF, Tang LJ, Wu YH, Cho BJ, Xu XJ, Dong JR, Song WD. Annealing and oxidation of silicon oxide films prepared by plasma-enhanced chemical vapor deposition. Journal of Applied Physics. 2005;**97**:014913

[25] Coyopol A, Cardona MA, Diaz-Becerril T, Licea-Jiménez L, Morales-Sánchez A. Silicon excess and thermal annealing effects on structural and optical properties of co-sputtered SRO films. Journal of Luminescence. 2016;**176**:40-46

[26] Yi LX, Heitmann J, Scholz R, Zacharias M. Si rings, Si clusters, and Si nanocrystals—different states of ultrathin SiOx layers. Applied Physics Letters. 2002;**81**:4248

[27] Coyopol A, García-Salgado G, Díaz-Becerril T, Juárez H, Rosendo E, López R, Pacio M, Luna-López JA, Carrillo-López. Optical and structural properties of silicon nanocrystals embedded in SiOx matrix obtained by HWCVD. Journal of Nanomaterials. 2012;**2012**(368268):7

[28] Coyopol A, Cabañas-Tay SA, Díaz-Becerril T, García-Salgado G, Palacios-Huerta L, Morales-Morales F, Morales-Sánchez A. Enhancement of the luminescence by the controlled growth of silicon nanocrystals in SRO/SiO_2 superlattices. Superlattices and Microstructures. 2017;**112**:534-540

[29] Pai PG, Chao SS, Takagi Y, Lucovsky G. Infrared spectroscopic study of SiOx films produced by plasma enhanced chemical vapor deposition. Journal of Vacuum Science & Technology A. 1986;**4**:689-694

[30] Zatryb G, Podhorodecki A, Misiewicz J, Cardin J, Gourbilleau F. Correlation between matrix structural order and compressive stress exerted on silicon nanocrystals embedded in silicon-rich silicon oxide. Nanoscale Research Letters. 2013;**8**:1-7

[31] Wolkin MV, Jorne J, Fauchet PM, Allan G, Delerue C. Electronic states and luminescence in porous silicon quantum dots: The role of oxygen. Physical Review Letters. 1999;**82**:197-200

[32] Khriachtchev L, Novikov S, Lahtinen J. Thermal annealing of Si/SiO_2 materials: Modification of structural and photoluminescence emission properties. Journal of Applied Physics. 2002;**92**:5856-5862

[33] Khriachtchev L, Rasanen M, Novikov S, Pavesi L. Systematic correlation between Raman spectra, photoluminescence intensity, and absorption coefficient of silica layerscontaining Si nanocrystals. Applied Physics Letters. 2004;**85**:1511-1513

Chapter 2

Nanograin Formation within Shear Bands in Cold-Rolled Titanium

Abstract

Microstructure evolution within the shear localization areas in a commercial titanium plate subjected to cold rolling was systematically investigated. A shear band with a width of approximately 25 μm was formed. The microstructure inside the shear band was mainly equiaxed nanograins with an average size of 70 nm. Transmission electron microscopy (TEM) observations revealed that the grain refinement inside the shear band was completely via a shear deformation-induced splitting and breakdown twin lamella process, instead of a nucleation and growth of new grains. The shear localization starts with the formation and multiplication of mechanical twins, which leads to the development of a twin/matrix lamellar structure aligned along the shear direction. The twin/matrix lamellae subsequently undergo gradual splitting and transverse breakdown, giving rise to fine elongated subgrains. The continuing lath breakdown, in combination with grain lateral sliding and lattice rotations, ultimately leads to the formation of a mix of roughly equiaxed, nanosized (sub)grains within the center of macroscopic shear band at large strains.

Keywords: nanograin, shear band, titanium, cold rolling, grain refining mechanism

1. Introduction

Shear band is a highly localized deformation mode that usually develops in a majority of metallic materials subjected to heavy loading, which generally leads to their limited applicability and high susceptibility to catastrophic failure in load-bearing applications [1, 2]. In the past two decades, the residual microstructures within shear bands have been studied at large using transmission electron microscopy, facilitated by the band convenient location in a restricted area of the hat-shaped specimens used in dynamic impact experiments [3–16]. These shear bands are frequently termed "adiabatic" shear bands, as they usually experience very high levels of strain rate, and the shear localization is frequently considered to be adiabatic with large temperature increases inside the bands [17]. The microstructures within adiabatic shear bands often undergo subsequent modifications through the recovery, recrystallization and phase transformation processes, which generally contribute to significant microstructure refinement through the formation of (sub)grains/fragments with the size of several tens of nanometers [1, 10].

The shear bands occurring in metals deformed by rolling and other quasi-static deformation modes [18–27] generally display microstructural features that markedly differ from those observed within their adiabatic counterparts. Strong texturing as well as elongated and heavily dislocated grains within the shear bands in a consolidated ultrafine-grained iron subjected to quasi-static compression has been revealed [27], which is strikingly different from the microstructure commonly found in the center of adiabatic shear bands. Refs. [18, 28, 29] performed detailed investigations, using a combination of the electron backscattering diffraction (EBSD) in a scanning electron microscope and the transmission electron microscopy (TEM) techniques, on the evolution of relatively coarse and/or well-recovered microstructures within shear bands formed during rolling or plane strain compression. At present, however, such investigations on the microstructures with extremely fine grains and high dislocation densities produced by heavy rolling deformation at room temperature are rather scarce and significantly less detailed. Such investigation is thus rather difficult and requires the time-consuming TEM technique. Furthermore, the rather random distribution of these bands within the matrix also causes the difficulty in preparation of targeted TEM specimens. The scale and character of the microstructures observed inside these shear bands have been found to share some similarity to those observed in ultrafine-grained materials processed by severe plastic deformation (SPD) [1, 30]. The microstructure refining mechanisms operating within the shear bands subjected to heavy rolling deformation will help to the better understand the grain-refining mechanisms occurring in other SPD processes. Nevertheless, deformation mechanisms themselves have frequently been able to achieve a significant refinement of the adiabatic shear band microstructure, in particular when these mechanisms involved mechanical twinning.

Apart from dislocation slip processes, mechanical twinning plays an important role in plastic deformation of hexagonal close packed (HCP) structure metals that have a limited number of slip systems. The predominant twinning systems are $\{10\bar{1}2\}\langle 10\bar{1}\bar{1}\rangle$ tensile twins and $\{11\bar{2}2\}\langle 11\bar{2}\bar{3}\rangle$ compression twins during deformation of hcp titanium at ambient temperature, accommodating extension and contraction along the c-axis, respectively [31, 32]. The intersection of twins and the formation of secondary and tertiary twins result in progressive grain refinement. At relatively modest strains, a gradual decrease in twin activity and saturation in twinning ultimately have been achieved, causing dislocation slip to dominate the deformation process at high strains [32–35]. The extremely fine, roughly equiaxed grains with a mean size of 80–100 nm in commercial purity titanium subjected to accumulated roll bonding has been reported [36]. These grains might possibly initiate from macro-shear and micro-shear bands, but details of their formation mechanism still remain to be elucidated.

This chapter will review our work on the nanograin formation in a cold-rolled titanium and supplement and update the extensive article published by our group. We performed a detailed TEM study on the microstructure evolution within the shear bands in commercial purity titanium subjected to heavy cold rolling and focus on elucidating the microstructure-refining mechanisms within the shear bands.

2. Experimental procedures

A titanium plate with a mean grain size of about 60 μm was cold rolled from 12 to 2 mm in thickness with a per pass reduction of 16.7% at a strain rate of 3 s^{-1}. The von Mises equivalent strains corresponding to different rolling reductions were calculated by $\varepsilon_{VM} = \frac{-2}{\sqrt{3}}\ln\left(\frac{t}{t_0}\right)$, with t_0 and t being the plate thickness before and after rolling, respectively. The shear strain accumulated within the shear bands was estimated using the method introduced by Ref. [9]. The microstructure both outside and within the shear localization areas was investigated by scanning electron microscopy (SEM) and TEM. An SEM study was carried out using a Zeiss Supra 55VP field-emission gun microscope operated at 10 kV. TEM investigation was performed using a Jeol JEM 2100 LaB_6 microscope operated at 200 kV. The observation sections were perpendicular to the transverse direction (TD) of the rolled plate. The etching of the samples was carried out using a solution composed of 50 ml H_2O, 40 ml HNO_3, and 10 ml HF.

The TEM samples are prepared as follows. A slice of the rolled specimen was first cut perpendicular to TD. In order to locate the shear bands, one side of the slice was metallographically polished and etched, then a light scratch was used to mark the shear band location and the rolling direction. The sample slice was then thinned by grinding from the opposite side to a sheet with thickness of approximately 100 μm. A disc with a diameter of 3 mm was subsequently

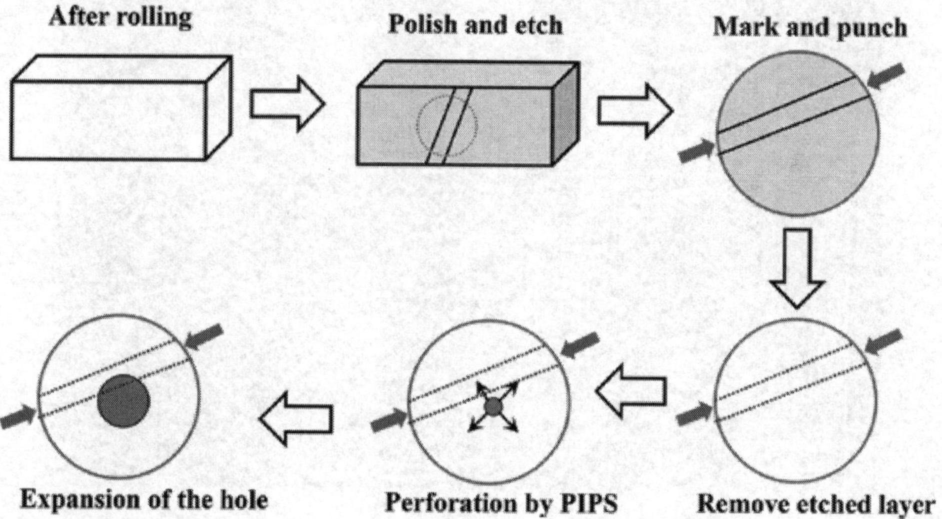

Figure 1. Schematic illustration of the preparation process for the TEM samples.

punched out from the sheet, ensuring that the intersection of the two marker lines was located close to the center of the disc, and the rolling direction was marked through a pair of fine notches placed on the disc rim. The disc was then carefully ground to approximately 50 μm in thickness and finally subjected to perforation using a low-energy ion milling in a Gatan PIPS system. **Figure 1** is a schematic illustration of the process for the preparation of TEM specimen.

3. Results

3.1. SEM observation of shear localization

Figure 2 shows that the micro-regions with localized shear were first initiated close to the edge of the rolled plate at von Mises equivalent strains (εVM) of 0.47 (rolling reduction of 33%) (**Figure 2a**). The S-like flow lines were extensively stretched in these micro-regions. The surrounding elongated grains, however, contained extensive deformation twinning, as shown in **Figure 3**. After 50% rolling reduction (εVM = 0.80), a localized microscopic shear band, having a width of approximately 8 μm and being inclined at an angle of about 40°relative to the rolling direction, was formed (**Figure 2b**); the deformation twins and the stretched grain boundaries located at the shear band boundaries gradually bent in a curve appearance when

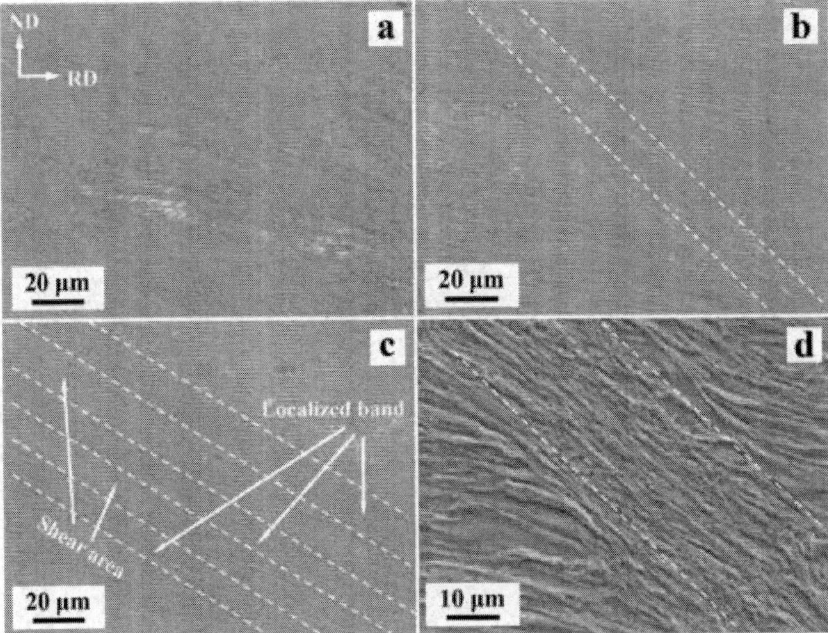

Figure 2. SEM micrographs showing the shear bands (delineated by dashed lines) developed at different rolling reductions: (a) 33% (εVM = 0.47), (b) 50% (εVM = 0.80), (c) 67% (εVM = 1.27), and (d) 83% (εVM = 2.07). RD and ND indicate the rolling and normal directions, respectively.

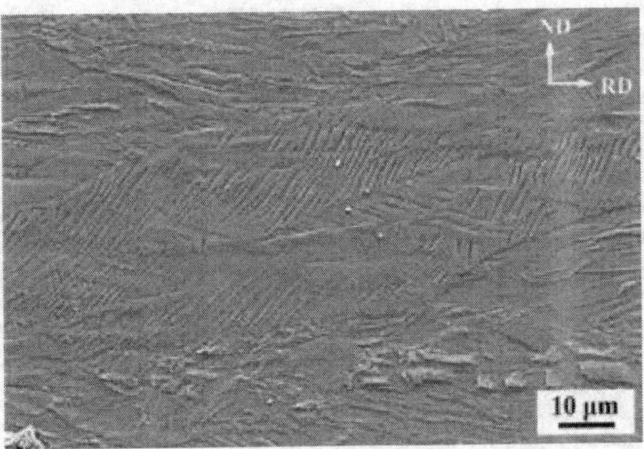

Figure 3. SEM micrograph showing extensive deformation twinning within the deformed grains at 33% rolling reduction (εVM = 0.47).

merging into the band. At the rolling reduction of 67% (εVM = 1.27), several microscopic shear bands formed and widened by expensing the adjacent deformed matrix regions (**Figure 2c**). When the rolling reduction reached 83% (εVM = 2.07), a well-developed macroscopic shear band having a width of about 25 μm was formed and the entire sheared region was filled with parallel flow lines (**Figure 2d**).

3.2. TEM quantification of the microstructure within shear bands

After 33% rolling reduction, the sheared micro-regions showed a significantly finer substructure than the surrounding matrix (**Figure 4a**). The selected area diffraction (SAD) patterns obtained from the micro-regions (inset in **Figure 4b**) indicated the occurrence of large-angle misorientation fragments. The neighboring matrix mainly consisted of dislocation cells separated by low-angle boundaries with a significant dislocation density in the interior. The corresponding SAD patterns (the inset in **Figure 4a**) were close to those typically obtained for single crystals. The sheared micro-regions contained fine twin/matrix lamellae, thin laths and elongated subgrain structures interspersed with the deformed matrix comprising dislocation cells. The elongated subgrains, however, could not be clearly resolved due to a high density of dislocations (**Figure 4b**).

When increased to 50% rolling reduction, a localized microscopic shear band contained regions of thin lamellae/laths interspersed with fine elongated subgrains (**Figure 5a**). **Figure 5b** shows that fine elongated subgrains contained lower dislocation density in the interior which are more evident than those formed at 33% rolling reduction. The histogram presented in **Figure 5c** shows a rather broad subgrain size distribution of 20–240 nm with a mean value of about 110 nm. Statistical determination of the subgrain longitudinal (d_L) and transverse length (d_T) revealed that the average d_L/d_T ratio was about 2.5 (inset in **Figure 5c**). Statistical evaluation of θ, the angle between the subgrain elongation axis and the rolling direction (RD), has an average value of around 40°, indicating a strong morphological texture with most of the subgrains retaining their elongation axis parallel to the shear direction (**Figure 5d**).

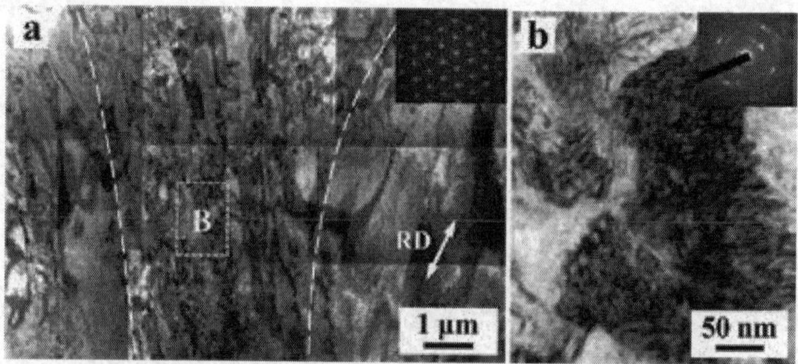

Figure 4. TEM micrographs obtained after 33% rolling reduction ($\varepsilon VM = 0.47$): (a) bright-field image of a sheared micro-region (delineated by dashed lines) and the matrix. The inset shows the matrix SAD pattern and RD indicates the rolling direction; (b) high-magnification bright-field image of the sheared micro-region area marked in (a) (the inset shows the corresponding SAD pattern).

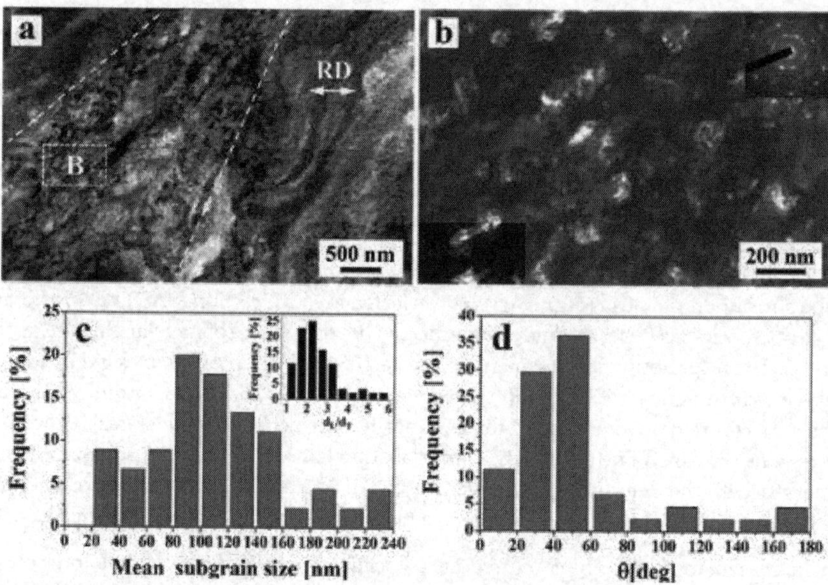

Figure 5. TEM micrographs obtained after 50% rolling reduction ($\varepsilon VM = 0.80$): (a) bright-field image of a region containing the localized microscopic shear band delineated by dashed lines. RD indicates the rolling direction; (b) dark-field image of the shear band area marked in (a) (the inset shows the corresponding SAD pattern); (c) subgrain size distribution for the microscopic shear band (the inset shows the corresponding histogram of dL/dT ratios); (d) distribution of the angles between the subgrain elongation axes and RD.

At 67% rolling reduction, the microstructure inside the localized microscopic shear bands became more homogeneous but maintained its alignment parallel to the shear direction (**Figure 6a**). The very fine, elongated rectangular or elliptical subgrains were observed (**Figure 6b** and **c**).

The SAD pattern of the corresponding area (inset in **Figure 6c**) exhibits diffuse arcing spots indicating large misorientations between adjacent subgrains. **Figure 6d** shows the histogram, indicating that most of the subgrains had a size in the range of 20–200 nm with a mean value of about 90 nm. The corresponding average value of the d_L/d_T ratio was about 2.0 (inset in **Figure 6d**). The distribution of θ (**Figure 6e**) indicates that the fine subgrains had a more random morphological alignment than those at 50% reduction.

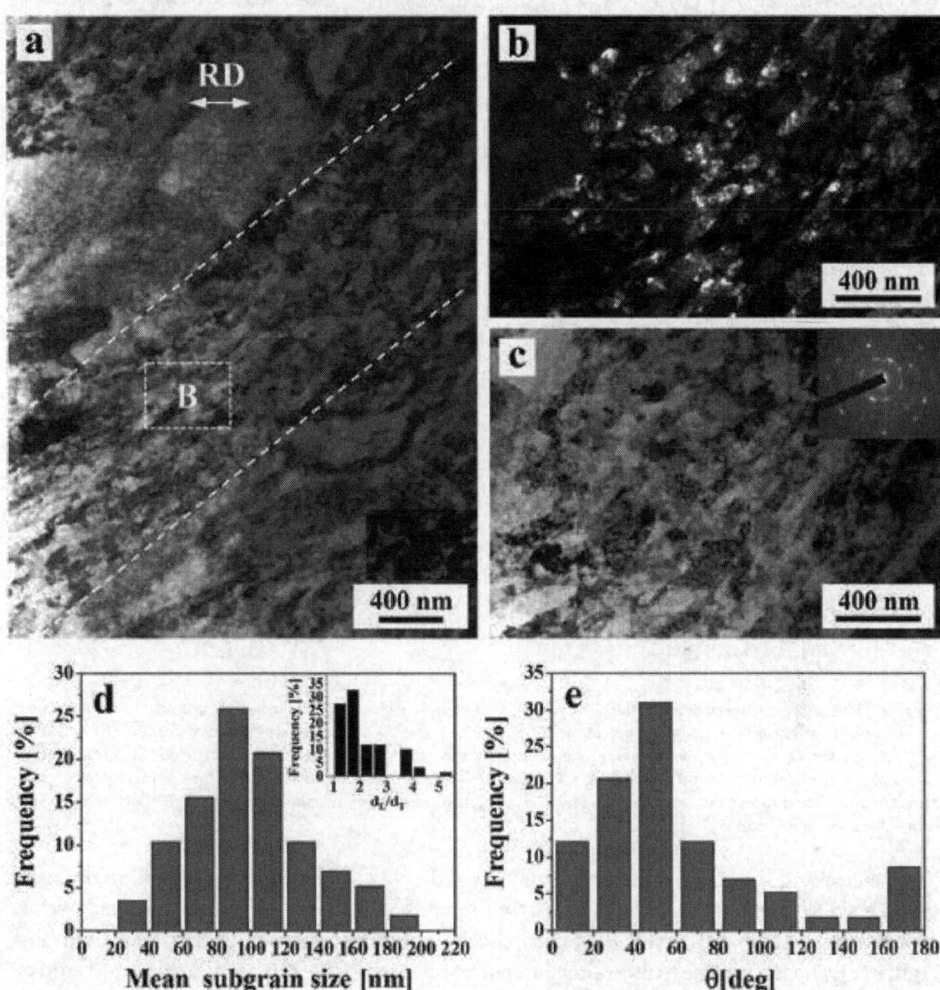

Figure 6. TEM micrographs obtained after 67% rolling reduction: (εVM = 1.27) (a) bright-field image of a region containing the localized microscopic shear band delineated by dashed lines. RD indicates the rolling direction; (b) dark-field image of the subgrains in the shear band area marked in (a); (c) bright-field image of the same area (the inset shows the corresponding SAD pattern); (d) subgrain size distribution for the microscopic shear band (the inset shows the corresponding histogram of dL/dT ratios); (e) distribution of the angles between the subgrain elongation axes and RD.

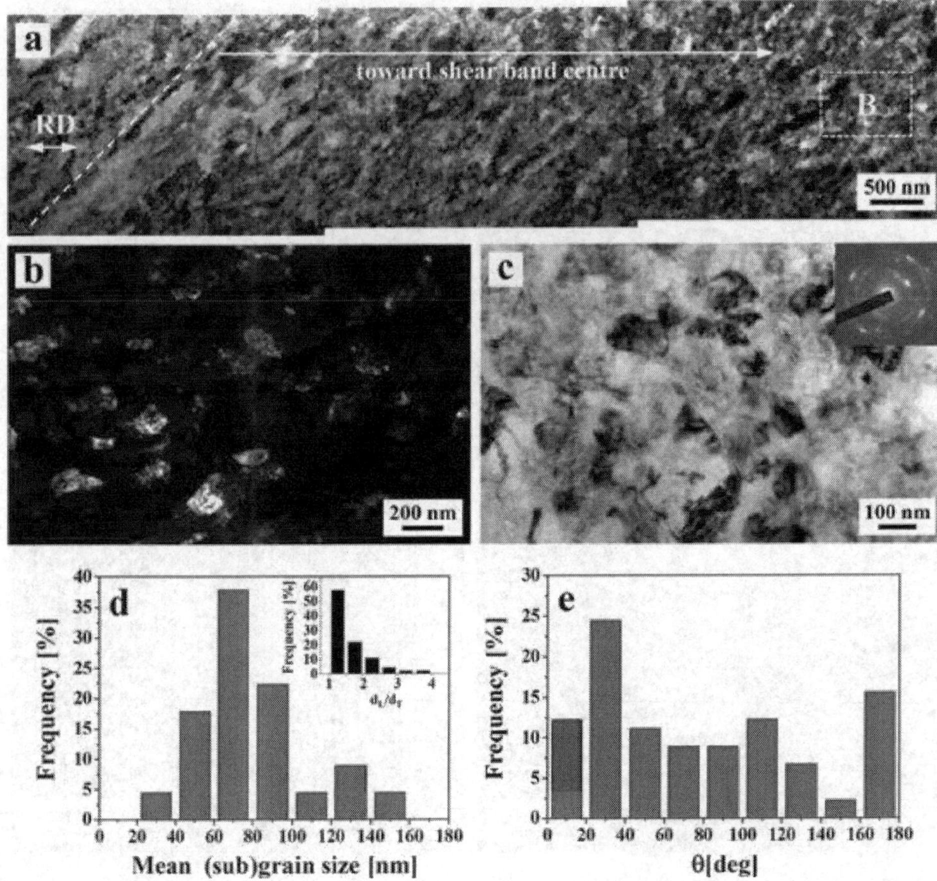

Figure 7. TEM micrographs obtained after 83% rolling reduction (εVM = 2.07): (a) bright-field image of a part of the macroscopic shear band with its boundary marked by a dashed line. RD indicates the rolling direction; (b) dark-field image of the nanosized (sub)grains present in the shear band central region indicated in (a); (c) bright-field image of the same area (the inset shows the corresponding SAD pattern); (d) (sub)grain size distribution for the macroscopic shear band center (the inset shows the corresponding histogram of d_L/d_T ratios); (e) distribution of the angles between the (sub) grain elongation axes and RD.

When the rolling reduction increased to 83%, two distinct regions developed within the macroscopic shear band. Wide outer regions were mainly filled with fine elongated subgrains, whereas the shear band center was occupied by roughly equiaxed nanosized (sub)grains (**Figure 7a**). These ultrafine (sub)grains showed no trace of the shear direction (**Figure 7b** and **c**). Some of these (sub)grains appeared to contain few dislocations and were delineated by rather sharp boundaries. **Figure 7d** reveals that the (sub)grains possessed a size range from 20 to 160 nm and had a mean size of about 70 nm. The corresponding average value of the d_L/d_T ratio was approximately 1.2 (inset in **Figure 7d**), clearly showing that these nanosized (sub) grains were mostly roughly equiaxed. The distribution of θ became randomized (**Figure 7e**), indicating the non-contiguous nature of these nanosized (sub)grains.

3.3. The microstructure refinement mechanisms within shear bands

The boundary regions between the shear localization areas and the matrix, as well as the outer regions of the localized bands, provided important information of the mechanism of microstructure evolution within the shear bands. **Figure 8** shows the mechanical twin in the boundary between a sheared micro-region and the matrix at 33% rolling reduction. The twin lamella consisted of several segments, three of which are labeled T_1, T_2 and T_3 in **Figure 8a**. It should be noted that the TEM foil was tilted so that the twin boundary displayed a minimum projected width and thus was in the "edge-on" position. The SAD patterns for the surrounding matrix (**Figure 8b**) and the main part of the twin lamella T_1 (**Figure 8c**) were both close to the $[1\bar{2}10]$ zone axis. The precise zone axes for the matrix and twin were reached at slightly different foil tilts by about 2°. These patterns shared a coincidental $(10\bar{1}2)$ reflection (circled in **Figure 8b** and **c**) and the corresponding diffraction spots displayed mirror symmetry with respect to a common $(10\bar{1}2)$ plane, which was approximately parallel to the twin planes (**Figure 8d**). This indicates that the twin segment T_1 represented a fine $\{10\bar{1}2\}\langle10\bar{1}\bar{1}\rangle$ tensile twin, which was further

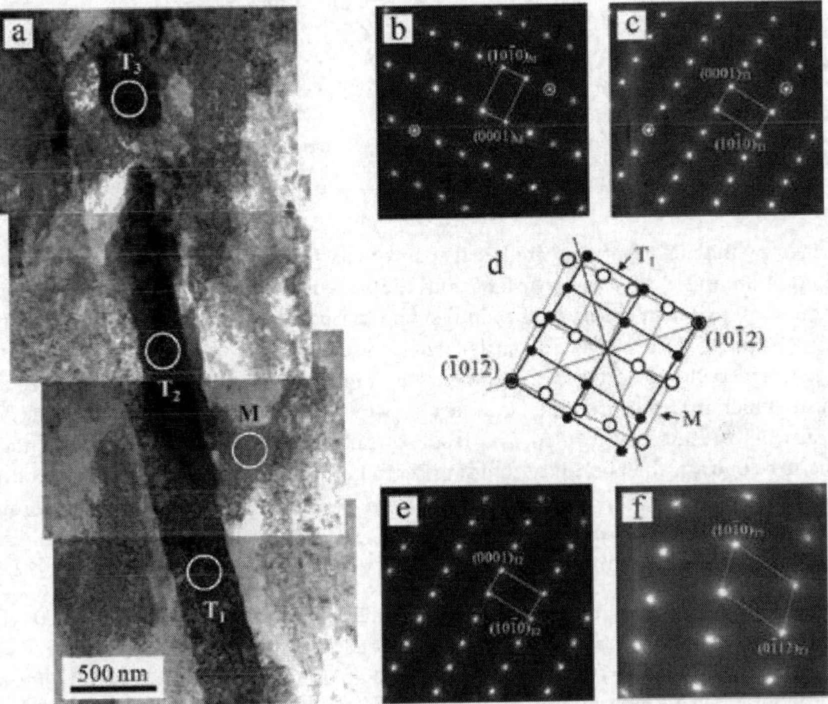

Figure 8. TEM analysis of a mechanical twin formed in the area separating a sheared micro-region and the matrix at 33% rolling reduction (ε_{VM} = 0.47): (a) bright-field image of the twin composed of segments T_1, T_2 and T_3 and embedded in the matrix M; (b), (c), (e), (f) SAD patterns obtained from the M, T_1, T_2 and T_3 regions, respectively (the corresponding locations are indicated by circles in (a)). The zone axis for the SAD patterns in (b), (c), (e) is $[1\bar{2}10]$ and for the SAD pattern in (f) this axis is $[\bar{2}4\bar{2}3]$; (d) schematic of the superimposed reciprocal lattice sections corresponding to the SAD patterns in (b) and (c) (the solid line indicates the $(10\bar{1}2)$ twinning plane).

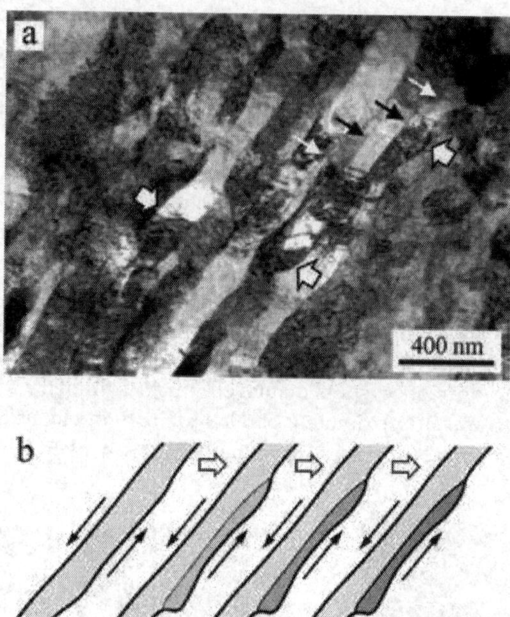

Figure 9. (a) TEM bright-field micrograph of the thin lath structure formed in a boundary region of the macroscopic shear band after 83% rolling reduction (ε_{VM} = 2.07); (b) schematic representation of the lamella longitudinal splitting process enhanced by shear induced bulging of its boundary (after ref. [10]), see text for details.

confirmed by the misorientation angle/axis pair of ~85°/ $[1\bar{2}10]$ derived from the SAD patterns (see **Figure 8b** and **c**), that is consistent with that calculated for the above twin system [31, 32]. Zhu et al. [35] also observed the similar fine $\{10\bar{1}2\}\langle 10\bar{1}\bar{1}\rangle$ twins in the microstructure of a commercial purity Ti processed by surface mechanical attrition treatment (SMAT). The SAD patterns corresponding to the twin segments T_1 (**Figure 8c**) and T_2 (**Figure 8e**) were almost identical, which indicates that the boundary separating these segments were low-angle dislocation wall, presumably formed as a result of deformation-induced splitting of the twin lamella. By contrast, the boundary dividing segment T_1 and T_3 was a high-angle boundary, as indicated by the respective SAD patterns (**Figure 8c** and **f**), suggesting that the segment T_3 might originate from the part of the twin lamella formation process. A similar subdivision of mechanical twin lamellae into high-angle boundary segments was also reported in [32, 37].

In the shear-localization regions, groups of fragmented mechanical twins interspersed with the matrix, elongated lamellar structure aligned along the shear direction were frequently observed. The elongated twin/matrix lamellae in the sheared areas seem to split through the formation of dislocation walls to form a thin lath microstructure with increasing shear strain. **Figure 9a** provides the evidence for the early stages of this process observed in the boundary region of the macroscopic shear band at 83% rolling reduction. An array of parallel elongated high-angle boundary laths containing some low-angle longitudinal dislocation walls was observed (indicated by small arrows in **Figure 9a**). It is clear that the twin/matrix lamellar structure

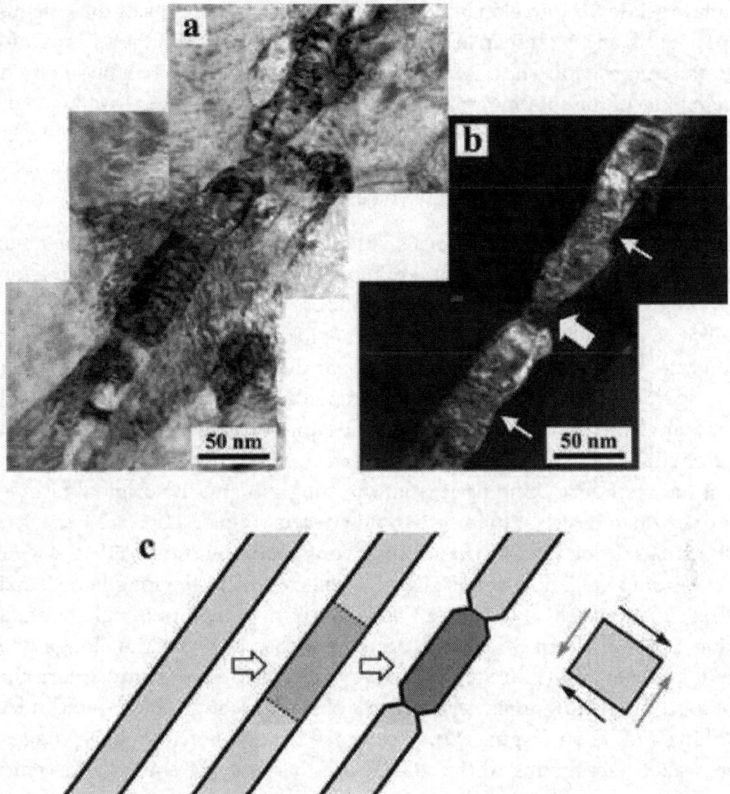

Figure 10. (a) TEM bright-field image of long laths breaking down into subgrains in a boundary region of the localized microscopic shear band after 67% rolling reduction (ε_{VM} = 1.27); (b) corresponding dark-field image; (c) schematic illustration of the breakdown process.

was progressively split and narrowed by the formation of such walls. With localized shear deformation progressing, the lath boundaries became curved and some lath regions appear to have locally been extruded out to form a bulge (marked by large arrows in **Figure 9a**). It was suggested in [10] that the bulge formation tends to accelerate the splitting process (**Figure 9b**). For a given rolling reduction, the observed range of the lath widths was generally consistent with the transverse lengths (d_T) of the fine elongated subgrains found within the shear bands. The above experimental observations thus suggest that these fine subgrains have likely formed through the deformation-induced transverse breakdown of the thin lath.

Figure 10 shows a typical long thin lath in the boundary region of a localized microscopic shear band at 67% rolling reduction. As shown, dislocations accumulated at several locations form "bamboo nodes" transverse dislocation (marked by the arrows in **Figure 10b**). The laths displayed a tendency to become constricted at the locations of the transverse boundaries (a clear example is indicated by the large arrow in **Figure 10b**). The formation of these walls led to the lath breakdown into elongated segments. The dimensions of these segments were about 30 nm in width and 100 nm in length, which is consistent with the sizes of the elongated subgrains in the microscopic shear band. This suggests that these bamboo-node dislocation walls are precursors of the subgrains and gradually became converted to elongated subgrains having large-angle boundaries with increasing shear strain. It should be noted that the laths in **Figure 10** also clearly experienced some splitting by longitudinal boundaries except for the breakdown through the formation of transverse dislocation walls.

Figure 11 shows a lath microstructure located at the localized microscopic shear band regions at 67% rolling reduction. It contains a subgrain being in the process of its formation and thus provides an insight into the mechanism of the gradual conversion of elongated lath segments to the fine elliptical, facetted subgrains found in regions of the macroscopic shear bands. The elliptical subgrain marked 2 in **Figure 11a** has clearly been initiated from a pre-existing segment of the long thin lath labeled 1. The subgrain, with few dislocations in the interior, is separated by high-angle boundaries from the neighboring areas marked 3 and 4 (**Figure 11c–e**), originating from the highly misoriented lath interfaces. The parent lath has become markedly constricted at the boundary dividing it from the subgrain and the original facet, separating regions 2 and 4, which has split to form two new facets (**Figure 11a**). TEM observation indicated that the subgrain has undergone a significant rotation (**Figure 11b** and **c**), evidenced by that the subgrain elongation axis is slightly deflected from the long lamella axis and the subgrain remains connected to the parent lath matrix by a medium-angle boundary with a misorientation angle of about 5°. Such types of linkages between the elongated subgrains and the parent lath matrix were frequently observed in the present study. Interestingly, some roughly equiaxed fine (sub)grains were already observed within the sheared micro-regions after 33% rolling reduction (**Figure 12**). The above observations strongly suggest that the equiaxed nanosized (sub)grains found within the macroscopic shear band center at large strains (**Figure 7**) did not originate from nucleation and growth process. Instead, they likely evolve through a continuous deformation-induced fragmentation of the pre-existing thin laths, first generating fine elongated subgrains followed by the formation of equiaxed fine (sub)grains with increasing shear strain.

Figure 13 displays an example of the microstructure formed in the regions near the macroscopic shear band at 83% rolling reduction. The microstructure was rather non-uniform and contained a mix of slightly elongated and roughly equiaxed fragments that clearly formed through a process of breaking down the pre-existing elongated laths. Two clearly discernible fragmented lath segments marked by dotted lines in **Figure 13** provide a further contribution to the understanding of the earlier fragmentation process. It is also seen from **Figure 13** that the microstructure was rather "turbulent," with the laths having different longitudinal axes and displaying pronounced local bending. This likely further improved the transverse lath breakdown. Furthermore, the equiaxed fragments boundaries were sharply discernible and the dislocation density within these fragments was relatively low. Some lateral sliding and bulging of the fragments out of the parent lath matrix, which generated new boundary facets, were also frequently observed (several examples are indicated by the arrows in **Figure 13**). The extended lath boundaries largely displayed medium-to-high misorientation angles while the misorientations across the shorter transverse boundaries ranged from low to high angles. Thus, most of

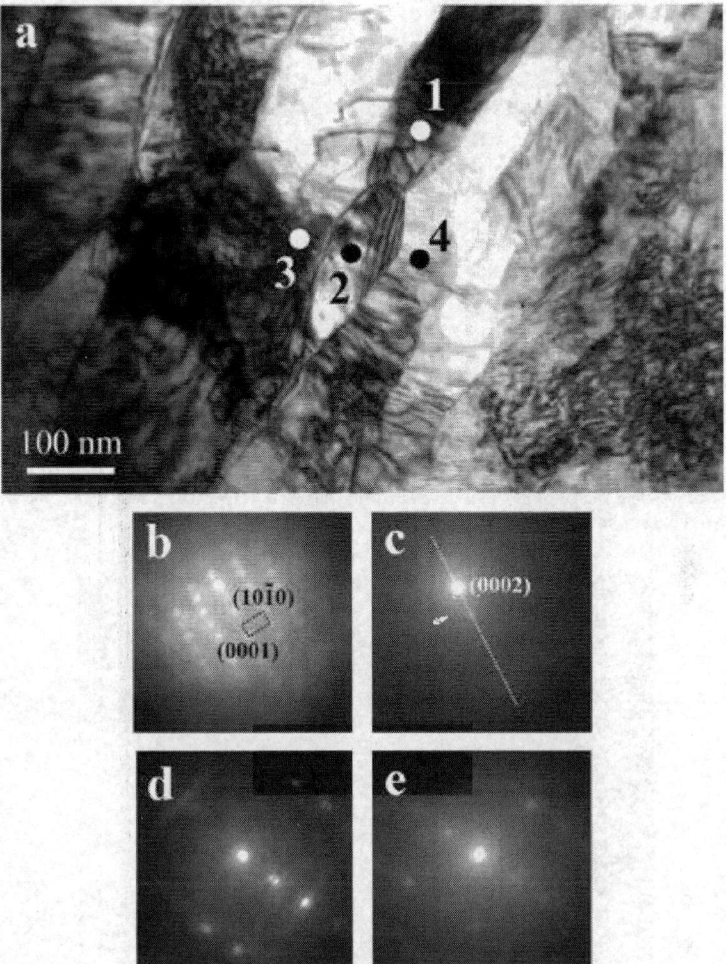

Figure 11. (a) TEM bright-field image of the thin lath structure at the boundary of the localized microscopic shear band after 67% rolling reduction (ε_{VM} = 1.27). The area contains a subgrain labeled 2 connected to the parent lath marked 1; (b-e) micro-diffraction patterns corresponding to the areas 1–4 labeled in (a), respectively. The diffraction pattern in (b) displays a [1$\bar{2}$10] zone axis and the pattern in (c) shows a systematic (0002) line of reflections (marked by the dotted line) resulting from a rotation of (b) by about 5° around the c-axis. Each of the diffraction patterns in (d) and (e) represents overlapping high-index zones and cannot be unambiguously indexed, but these patterns clearly show high-angle misorientations (above 15°) relative to (c).

the equiaxed fragments shown in **Figure 13** can be described as (sub)grains bounded partly by lower-angle and partly by high-angle boundary facets. Some of these fragments were completely bounded by high-angle boundary facets and can be thus classified as grains. The presence of polygonized, equiaxed nanosized (sub)grains, fully enclosed by high-angle boundaries, in the center, of shear bands was observed to be markedly higher than that found in the shear

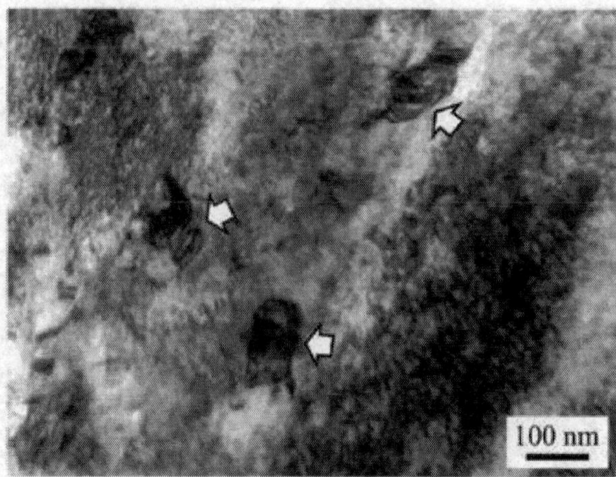

Figure 12. TEM bright-field image of the well-developed fine subgrains (arrowed) found within the sheared micro-regions after 33% rolling reduction ($\varepsilon_{VM} = 0.47$).

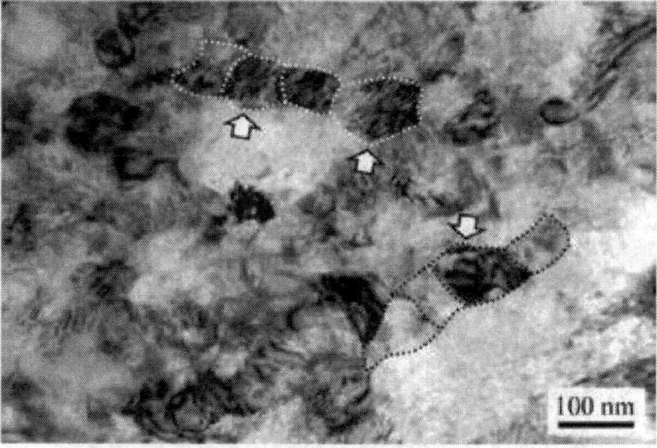

Figure 13. TEM bright-field image of the coexisting elongated subgrains and equiaxed (sub)grains in the macroscopic shear band outer region after 83% rolling reduction ($\varepsilon_{VM} = 2.07$).

band outer areas. This observation was likely a consequence of (sub)grain rotation promoted by a large localized shear strain developed in the shear band center, which in turn increased the frequency of high-angle boundaries compared to the outer shear band regions.

4. Discussion

The deformation-induced temperature increase plays an important role in the formation of the equiaxed fine (sub)grains inside the shear bands. Ref. [38] gives a simple calculation of the

Rolling reduction (%)	Von Mises strain ε_{VM}	Shear strain in the shear bands ε_s	Temperature increase in the matrix ΔT_M (K)	Temperature increase in the shear bands ΔT_S (K)
33	0.47	—	76	—
50	0.80	2.31	129	374
67	1.27	3.08	205	498
83	2.07	6.32	335	1020

Table 1. Estimated values of shear strain within the shear bands together with the deformation-induced temperature increases calculated for both the overall sample matrix and shear bands at different rolling reductions.

temperature increase during rolling deformation (ΔT): $\beta\sigma_{flow}\,\varepsilon = \rho c_p \Delta T$, where β is the thermal conversion factor taken as 0.9, σ_{flow} is the plane flow stress ($\sigma_{flow} = \frac{2}{\sqrt{3}} \times \sigma$ where σ is the tensile strength equal to 366 MPa), ε is the deformation strain, ϱ is the density equal to 4.5 g cm^{-3}, and c_p is the specific heat capacity taken as 523 J kg^{-1} K^{-1}. It is shown in **Table 1** that the temperature increase at the center of shear bands is markedly higher than that for the entire specimen. In the central region of the macroscopic shear band at 83% rolling reduction, the temperature increase estimated (~1020 K) is slightly higher than the expected recrystallization temperature of titanium [taken as 0.5 T_m, where T_m is the melting temperature of titanium (1943 K)]. However, during the rolling process, considerable specimen heat losses inevitably have occurred by both convection to the rolls and to air cooling. Therefore, the actual deformation-induced temperature increases are significantly lower than the values given in **Table 1**. This suggests that the recrystallization threshold has not likely been exceeded in the macroscopic shear band center areas, which is also supported by the absence of a microstructural feature form through the recrystallization mechanism in the shear band in our study. However, the significant temperature increases are expected to enhance the recovery processes.

The microstructure evolution within the shear localization areas can be summarized as follows. Micro-localized shear deformation regions, containing fine twin/matrix lamellae, thin laths and elongated subgrain structures interspersed with the deformed matrix, are first initiated at low strains. The formation and multiplication of approximately parallel, distinct microscopic shear bands inclined to the rolling direction at an angle of about 40° are motivated by further shear localization with increasing strain. The microscopic shear bands gradually coalesce to form a macroscopic shear band. The macroscopic shear band contains a complex structure composed of thin lath structures in the boundary regions, fine elongated subgrains in the outer areas and nanoscale, roughly equiaxed (sub)grains in the center. This suggests that the macroscopic shear band has a significant strain gradient, resulting from the band gradual development and the center region experiencing larger shear strain compared to the outer regions. The present microstructure investigation has thus clearly revealed that the above nanosized (sub)grains are formed entirely through a shear deformation-induced process. The roughly equiaxed grains in the center of shear band are a final product of the process of progressive splitting and breakdown of the thin elongated structures, which appear to partly originate from the matrix/twin lamellar structure aligned approximately parallel to the shear direction. Although there is no clear evidence of nucleation and growth of new recrystallized

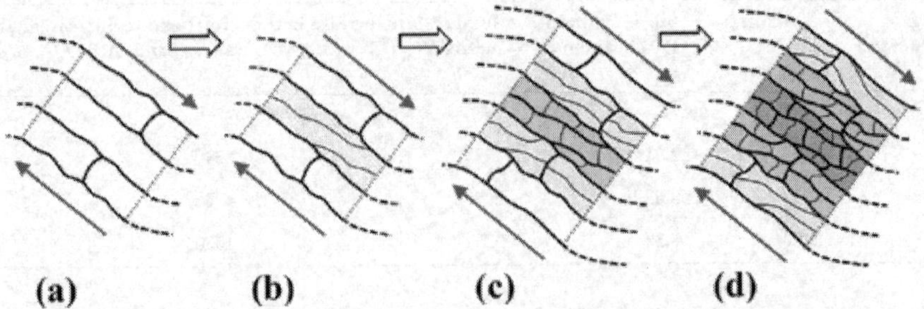

Figure 14. Schematic representation of the suggested mechanism of the microstructure evolution within the shear band interior with increasing strain: (a) formation of the mechanical twin/matrix lamellae; (b) longitudinal splitting of the lamellae to form thin laths; (c) transverse breakdown of the laths to form elongated subgrains; (d) further breakdown and rotation of the subgrains to form equiaxed nanoscale (sub)grains.

grains was observed, a deformation-induced temperature increase (**Table 1**) can be expected to enhance the recovery processes, in particular within the shear band center at large rolling reductions, which are consistent with the microstructure feature observed by TEM.

Figure 14 schematically illustrates the mechanism of the microstructure evolution within shear bands. Before the initiation of shear localization, mechanical twins are frequently observed within the deformed grains (**Figure 3**). These twins are often divided to several high-angle segments (**Figure 8**), presumably as a result of the multiple twinning events which have been reported to frequently occur within the mechanical twins formed in HCP materials [31, 32, 37]. The activation of additional twin systems is inevitably triggered by significant lattice rotation in the areas of shear localization. Then the twins are incorporated into the localized bands and gradually reorient themselves toward the shear band direction. These mechanical twins rapidly multiply until they reach a saturated state to form twin/matrix lamellar structure with a relatively narrow spacing (**Figure 14a**).

The longitudinal splitting and transverse breakdown are the two main deformation-induced processes which contribute to the gradual transition from the elongated twin/matrix lamellar structure to the final roughly equiaxed (sub)grains. The transition from the elongated twin/matrix lamellae to the thin lath structure (**Figure 14b**) is primarily achieved by longitudinal splitting mechanism. The conversion of the thin lath structure to the fine elongated subgrains is mainly controlled by the transverse breakdown mechanism (**Figure 14c** and **d**). Although longitudinal splitting generally occurs prior to intense transverse breakdown, these two mechanisms might occur simultaneously.

The mechanical twinning gradually is consumed and the dislocation activity then dominates the deformation process at large strains [35, 39]. At this stage, the progressive splitting, and narrowing, of the matrix/twin lamellae leads to the formation of finely spaced lath structure containing a mix of twin boundaries and dislocation walls (**Figure 7a**). With increasing strain, twin boundaries gradually become severely distorted and thus converted to general high-angle boundaries [40]. At same time, the dislocation walls finally become transformed to high-angle boundaries through continuing to absorb dislocations and increase their misorientations [30, 40]. As a result, it is frequently hard to clearly distinguish the origin of the lath boundaries at large strains. Ref. [10] suggested that the above splitting process is promoted by the interactions between the adjacent elongated lamellar segments under heavy shear deformation. Once the lamellae develop some perturbations or bulges, largely created due to their transverse breakdown into shorter segments, their longitudinal boundaries become undulating (**Figure 9a** and **10**). The bulges then split from the parent lamellae driven by the adjacent segments, which result in further narrowing of the lamellae and the formation of a thin lath structure (**Figure 9b**).

The dislocations within the laths tend to accumulate at some locations to form "bamboo nodes" transverse dislocation walls. This process first leads to the formation of fine elongated subgrains through the breakdown of long lath progressive (**Figures 5, 6** and **10–12**) and finally, in conjunction with lattice rotations, to the formation of roughly equiaxed fine (sub)grains (**Figures 7** and **13**). It has been suggested that the driving force for this process is mainly provided by the conjugated shear stress [10]. This shear stress is also likely to promote the observed lateral sliding of some of the lath segments out of the parent matrix, which leads to the formation of new large-angle boundary facets (**Figures 11–13**). This mechanism is thus expected to contribute substantially to the transition from the elongated lath subgrains to the roughly equiaxed nanosized (sub)grains found in the macroscopic shear band core region (**Figure 7**). It seems plausible to expect that, with further increase in localized shear strain, the fragmentation and rotation processes within the macroscopic shear band center would ultimately lead to the formation of equiaxed nanosized grains fully enclosed by high-angle boundaries.

In the present chapter, isolated dislocations are frequently observed within both the fine elongated subgrains and ultrafine equiaxed (sub)grains. It is suggested that plastic deformation within these structures largely occurs through dislocation slip processes down to the smallest fragments observed. This is in agreement with the published data suggesting that plastic deformation was governed by the dislocation processes within microstructural fragments with 5 nm dimension [30, 40]. The gradual subdivision process of the starting matrix/twin lamellar structure of HCP titanium observed in our study within the shear localization areas might thus be principally described within the Risø framework of deformation microstructure evolution by slip processes [40], originally developed predominantly for FCC and BCC metallic materials. According to this concept, the matrix/twin lamellae tend to progressively subdivide during straining into cell blocks separated by "geometrically necessary" boundaries (GNBs) and containing "incidental" dislocation boundaries (IDBs) formed by statistical trapping of slip dislocations. Typical shapes of cell blocks accommodating large plastic strains are thin laths, delineated by extended GNBs and subdivided into shorter segments by a mix of transverse GNBs and IDBs. The lath structure parameters have been observed to scale in proportion to strain up to very large strain levels [40]. GNBs are predominantly large-angle boundaries while IDBs are generally low-angle dislocation walls at medium-to-large strains, and the misorientation angle for both the boundary types typically increases and their spacing decreases with increasing strain due to a continuous accumulation of dislocations in the structure. As a result, the GNBs are first converted into high-angle boundaries, followed by the conversion of transverse IDBs, and the grain subdivision process thus ultimately leads to the formation of nanosized equiaxed grains separated by high-angle boundaries at very large strains [40].

Terada et al. [36] reported the fine equiaxed grains of 80–100 nm in commercial purity titanium subjected to accumulated roll bonding. This mean grain size is just slightly higher than what has been achieved in the center of shear bands after 83% rolling reduction. This implies that these fine equiaxed grains might have indeed originated from macro-shear and micro-shear bands as tentatively suggested by the above authors. A considerable microstructure refinement generally associated with the adiabatic shear band development during dynamic loading was observed during the deformation by heavy cold rolling [11, 41]. The microstructure feature observed within the shear bands produced by cold rolling appears to bear some similarity to the microstructures typically obtained by SPD processes [1, 30]. It has been suggested that intense plastic deformation within the shear band might actually be classified as an SPD process [4]. Thus, it seems plausible that some of the microstructure refinement mechanisms suggested in the present chapter might be also applicable to the SPD grain-refining processes.

5. Conclusions

This chapter presents a detailed investigation on the microstructure evolution and nanograin formation within the shear localization areas formed in commercial purity titanium during cold rolling deformation. The grain-refining mechanism has been addressed and the following conclusions can be drawn:

1. Sheared micro-regions, containing fine twin/matrix lamellae, thin laths and elongated subgrains, are first initiated at low strains and further shear localization with increasing strain leads to the formation and multiplication of distinct microscopic shear bands. The microscopic shear bands gradually coalesce to form a macroscopic shear band. The macroscopic shear band contain a mix microstructure of thin lath structures, fine elongated subgrains and roughly equiaxed (sub)grains with a mean size of about 70 nm.
2. The shear localization starts from the formation and multiplication of twin/matrix lamellar structure aligned along the shear direction. The twin/matrix lamella then splits into thin laths through the formation of longitudinal dislocation walls. The thin laths gradually transverse breakdown by the formation of short transverse dislocation boundaries. The continuing thermally assisted lath breakdown, in conjunction with lateral sliding and lattice rotations, leads to the formation of a mix of roughly equiaxed, nanosized (sub)grains and grains in the macroscopic shear band center at large strains.
3. Some of the mechanisms of the microstructure refinement within the shear localization areas suggested in the present chapter might be also applicable to other severe plastic deformation processes. The gradual microstructure fragmentation process within the shear localization areas might be principally described within the Risø framework of deformation microstructure evolution by slip processes, originally developed predominantly for FCC and BCC metallic materials.

References

[1] Xu YB, Zhang JH, Bai YL. Shear Localization in Dynamic Deformation: Microstructural Evolution. Metallurgical and Materials Transactions A. 2008;**39A**:811

[2] Walley SM. Shear Localization: A Historical Overview. Metallurgical and Materials Transactions A: Physical Metallurgy andMaterials Science. 2007;**38A**:2629

[3] Kad BK, Gebert JM, Pérez-Prado MT, Kassner ME, Mayers MA. Ultrafine-grain-sized zirconium by dynamic deformation. Acta Materialia. 2006;**54**:4111

[4] Meyers MA, Xu YB, Xue Q, Pérez-Prado MT, McNelley TR. Microstructural evolution in adiabatic shear localization in stainless steel. Acta Materialia. 2003;**51**:1307

[5] Meyers MA, Cao BY, Nesterenko, Benson D. Xu YB. Shear localization-martensitic transformation interactions in Fe-Cr-Ni monocrystal. Metallurgical and Materials Transactions A: Physical Metallurgy and Materials Science. 2004;**35A**:2575

[6] Andrade UR, Meyers MA, Vecchio KS, Chokshi AH. Dynamic recrystallization and grain size effects in shock hardened copper. Acta Metallurgica et Materialia. 1994;**17**:175

[7] Pérez-Prado MT, Hines JA, Vecchio KS. Micmstructural evolutioninadiabaticshearbandsin TaandTa-Walloys {J}. Acta Materialia. 2001;**49**:2905

[8] Xue Q, Cerreta EK, Gray GT III. Microstructural characteristics of post-shear localization in cold-rolled 316L stainless steel. Acta Materialia. 2006;**55**:691

[9] Xue Q, Gray GT III. Development of adiabatic shear bands in annealed 316L stainless steel: Part I. Correlation between evolving microstructure and mechanical behavior. Metallurgical and Materials Transactions A: Physical Metallurgy and Materials Science. 2006;**37**:2435

[10] Xue Q, Gray GT III. Development of adiabatic shear bands in annealed 316L stainless steel: Part I. Correlation between evolving microstructure and mechanical behavior. Metallurgical and Materials Transactions A: Physical Metallurgy and Materials Science. 2006;**37**:2447

[11] Chichili DR, Ramesh KT, Hemker KJ. Adiabatic shear localization in α. Journal of the Mechanics and Physics of Solids. 2004;**52**:1889

[12] Xu YB, Bai YL, Shen LT. Formation, microstructure and development of the localized shear deformation in low-carbon steels. Acta Materialia. 1996;**44**:1917

[13] Wittman CL, Mayers MA, Pak H-r. Metallurgical Transactions A. Observation of an adiabatic shear band in AISI 4340 steel by high-voltage transmission electron microscopy. 1990;**21**:707

[14] Timothy SP, Hutchings IM. The structure or adiabatic shear bands in a titanium alloy. Acta Metallurgica. 1985;**33**:667

[15] Timothy SP. Acta Metallurgica. The structure of adiabatic shear bands in metals: A critica l review. 1987;**35**:301

[16] Wei Q, Kecskes L, Jiao T, Hartwig KT, Ramesh KT, Ma E. Adiabatic shear banding in ultra-fine-grained Fe processed by severe plastic deformation. Acta Materialia. 2004;**52**:1859

[17] Wright TW. The Physics and Mathematics of Adiabatic Shear Bands. Cambridge: Cambridge University Press; 2002

[18] Paul H, Morawiec A, Bouzy E, Fundenberger JJ, Piatkowski A. Brass-type shear bands and their influence on texture formation. Metallurgical and Materials Transactions A: Physical Metallurgy and Materials Science. 2004;**35A**:3775

[19] Dillamore IL, Roberts JG, Bush AC. Occurrence of shear bands in heavily rolled cubic metals. Metal Science. 1979;**2**:73

[20] Canova GR, Kocks UF, Stout MG. On the origin of shear bands in textured polycrystals. Scripta Metallurgica. 1984;**18**:437

[21] Lee CS, Hui WT, Duggan BJ. Macroscopic shear bands in cross-rolled α brass. Scripta Metallurgica et Materialia. 1990;**24**:757

[22] Sevillano JG, Houtte PV, Aernoudt E. The contribution of macroscopic shear bands to the rolling texture of FCC metals. Scripta Metallurgica. 1977;**11**:581

[23] Nakayama Y, Morii K. Microstructure and shear band formation in rolled single crystals of Al-Mg alloy. Acta Metallurgica. 1987;**35**:1747

[24] Embury JD, Korbel A, Raghunathan VS, Rys J. Shear band formation in cold rolled Cu-6% Al single crystals. Acta Metallurgica. 1984;**32**:1883

[25] Chowdhury SG, Das S, De PK. Cold rolling behaviour and textural evolution in AISI 316L austenitic stainless steel. Acta Materialia. 2005;**53**:3951

[26] Ohsaki S, Kato S, Tsuji N, Ohkubo T, Hono K. Bulk mechanical alloying of Cu–Ag and Cu/Zr two-phase microstructures by accumulative roll-bonding process. Acta Materialia. 2007;**55**:2885

[27] Wei Q, Jia D, Ramesh KT, Ma E. Evolution and microstructure of shear bands in nano-structured Fe. Applied Physics Letters. 2002;**81**:1240

[28] Cizek P, Bai F, Rainforth WM, Beynon JH. Fine structure of shear bands formed during hot deformation of two austenitic steels. Materials Transactions. 2004;**45**:2157

[29] Cizek P. Electron backscatter diffraction (ebsd) – The Method and its applications in materials science and engineering. Materials Science and Engineering A. 2002;**324**:214

[30] Lu K, Hansen N. Structural refinement and deformation mechanisms in nanostructured metals. Scripta Materialia. 2009;**60**:1033

[31] Chun YB, Yu SH, Semiatin SL, Hwang SK. Effect of deformation twinning on micro-structure and texture evolution during cold rolling of CP-titanium. Materials Science and Engineering A. 2005;**398**:209

[32] Stanford N, Carlson U, Barnett MR. Deformation Twinning and the Hall–Petch Relation in Commercial Purity Ti Metallurgical and Materials Transactions A: Physical Metallurgy and Materials Science. 2008;**39A**:934

[33] Lutjering G, Williams JC. Titanium. Engineering Materials and Processes. New York: Springer; 2003

[34] Shin DH, Kim I, Kim J, Kim YS, Semiatin SL. Microstructure development during equal-channel angular pressing of titanium. Acta Mater. 2003;**51**:983-996

[35] Zhu KY, Vassel A, Brisset F, et al. Nanostructure formation mechanism of α-titanium using SMAT [J]. Acta Mater. 2004;**52**(14):4101-4110

[36] Terada D, Inoue S, Tsuji N. Microstructure and mechanical properties of commercial purity titanium severely deformed by ARB process. Journal of Materials Science. 2007;**42**:1673

[37] Cizek P, Barnett MR. Characteristics of the contraction twins formed close to the fracture surface in Mg–3Al–1Zn alloy deformed in tension. Scripta Materialia. 2008;**59**:959

[38] Jiang L, Pérez-Prado MT, Gruber PA, Arzt E, Ruano OA, Kassner ME. Microstructure and mechanical properties of equiaxed ultrafine-grained Zr fabricated by accumulative roll bonding. Acta Materialia. 2008;**56**:1228

[39] Hughes DA, Hansen N. Deformation structure developing on fine scales. Philosophical Magazine. 2003;**83**:3871

[40] Hansen N. Metallurgical and Materials Transactions A. New discoveries in deformed metals. 2001;**32A**:2917

[41] Meyers MA, Subahash G, Kad BK, Prasad L. Mechanics of Materials. Evolution of Microstructure and Shear-Band Formation in a-hcp Titanium. Mechanics of Materials. 1994;**17**:175

Chapter 3

The Nanostructure Zeolites MFI-Type ZSM5

Abstract

The development of new porous solids begins to provide effective and original solutions to the problems of pollution and sustainable development, and the challenge is to discover new performance of these materials. Among these materials are zeolites; however, one type has retained the attention since its discovery in 1972; this is zeolite-type ZSM-5, because of their particular properties in many industrial processes. These porous solids are characterized on the atomic scale by the existence of pores, distributed regularly in the matter and are likely to accommodate in their structures, gases, liquids, and solids for trap or temporarily store them. Only the average porosity of these zeolites constitutes an obstacle to the catalysis of cumbersome molecules that are well branched. To deflect this inaccessibility factor and trying to find a solution to this steric hindrance, the attention of the researchers was focused on using the surface properties of ZSM-5. It is well known that the efficiency and selectivity of a porous catalyst depends on its textural and structural characteristics and more precisely on the number of locations active on the external surface and the number of locations accessible through the porous system. The nanocrystallinity in the field of zeolites can be defined as a situation in which the physicochemical properties are largely determined by a larger number of atoms in the outer limit of crystallite. In this chapter, the work of several researchers who synthesized the ZSM5 in the field of nanostructures is presented. We find that, despite the different methods of synthesis, however, the field of nanostructures ZSM5 has been achieved. Certain parameters such as the concentration of mineral agent, the concentration of the structuring agent, and the duration of aging have a direct influence on the crystal size of the zeolites obtained. Different characterizations were used to identify the purity and size of the nanocrystals ZSM5.

Keywords: nanostructures, zeolites, catalysis, heterogeneous catalysts, ZSM5

1. Introduction

The discovery of a new family of inorganic materials composed mainly of silicon, aluminum, and oxygen has not ceased to attract growing interest and wide investigations. In 1756, Crönstedt [1], a Swedish mineralogist, discovered stilbite. He attributed it the name of zeolite which comes from the Greek *zeo* boiling and *lithos* stone (the boiling stone). Subsequently, the name zeolite was assigned to a family of natural aluminosilicic minerals. Their important scientific success and their kaleidoscopic applications are mainly due to the increasing progress made in synthesis (**Figure 1**).

Zeolites are hydrated aluminosilicates of the general chemical formula [2]: $M^{n+}_{(x/n)}$, $(AlO^-_2)_x$, $(SiO_2)_y$, $(H_2O)_m$. Their properties depend in part on the value of the Si/Al ratio [3]. The structure of the zeolite crystals consists of an assembly of TO_4 tetrahedra (**Figure 2**).

The tetrahedron arrangement having two common oxygenated peaks leads to the formation of SBUs (secondary building units), which serve as reference patterns for classifying and describing the various structures of zeolites and their microporous networks [4, 5]; these structural types are made up of common genetic entities.

The secondary units are assembled in turn into polyhedra; this assemblage gives rise to the final structure. Each structural type obtained is assigned with a three-letter code according to the IUPAC nomenclature (International Union of Pure and Applied Chemistry).

It can be either an aluminum atom or a silicon atom. The arrangement of the tetrahedra in the three directions of space generates a large microporosity in which molecules will be able to adsorb and hence the name of molecular sieve given by Dr. José Walkimar Mesquita Carneiro (JWMC).

The zeolites are probably the only family that offers a great deal of structural and chemical diversity; several types are currently known and synthesized [6].

Figure 1. A natural zeolite "stilbite."

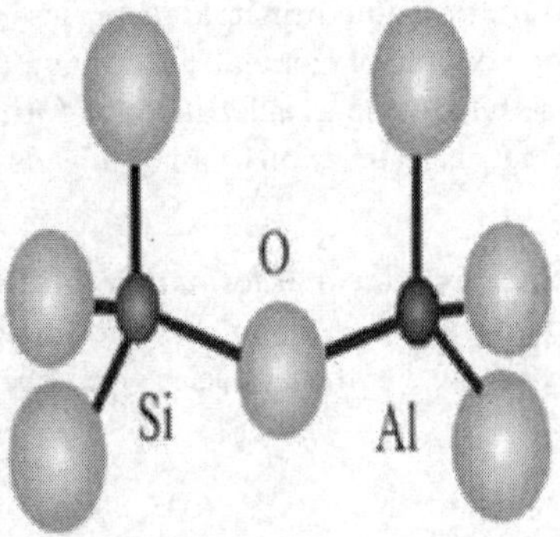

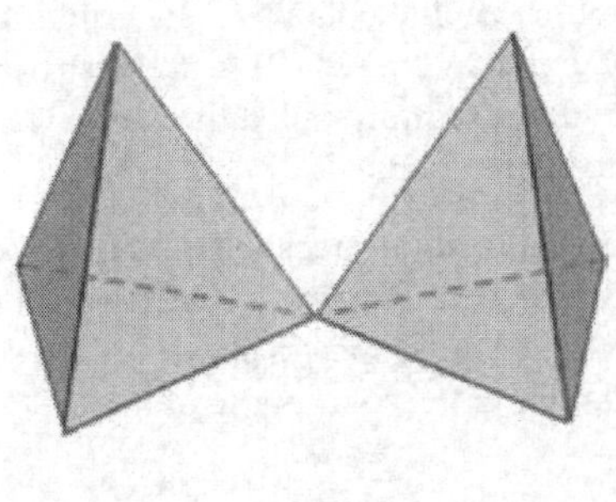

Figure 2. Diagram of the assembly of SiO_4 and AlO_4^- tetrahedra.

Their major element is a tetravalent element of silicon "Si"; it is usually accompanied by a trivalent element, aluminum "Al," iron "Fe," barium "Ba," and so on, and sometimes of a divalent element like beryllium "Be." The presence of elements in the framework with valence lower than four led to the appearance of the negative charge, which is compensated by cations M^{n+} (generally alkaline earth) mobile and interchangeable with other cations.

The natural zeolites were considered as early as 1862, but the zeolite synthesis was not made for the first time until 1956 [7]. Today, the zeolite family includes more than 190 natural or synthetic aluminosilicates [4], each one is characterized by its own porous structure. This microporosity offers them many properties in the fields of adsorption, purification, and molecular sieving in separation processes. The synthesis of these materials is now well controlled, and the introduction of the quaternary ammonium cation type, as amines into the reaction medium, has allowed the development of many microporous materials related to zeolites such as aluminophosphates or gallophosphates. So far, 194 structural types exist [8], whose name is designated, according to the structure commission IZA (International Zeolite Association), by a code of three capital letters [9]. The particular porous structure of the zeolites, with pores and size channels at the molecular level, was at the origin of their properties (molecular sieves, large specific surface area, etc.). In addition, most of these materials possess an interesting thermal and mechanical stability compatible with industrial applications in various fields.

Their most important applications remain in the field of catalysis [10]. The zeolites are also used for their acidic and/or redox properties and their high regeneration capacity in petroleum refineries, for the cracking of heavy hydrocarbons in gasoline. The zeolites as faujasite (X or Y), beta, and ZSM-5 are the three most frequently used materials.

The MFI (Mobil-type five) zeolites [11] chosen as model adsorbents in this chapter are zeolites ZSM-5 (Zeolite Socony Mobil-type 5) [12]; they were synthesized for the first time in 1972 [13]. These materials are thermally stable up to 1000°C and their organophilic characters (due to Si-O bonds) and hydrophobic (due to their low content of charge-compensating cation) make them adsorbents and catalysts of choice.

The preparation methods of these zeolites are based on the same principle regardless of the type of the zeolite obtained. The bringing together of a source of element T, a solvent (H_2O), a mobilizing agent (OH^-, F^-), and a structuring and stabilizing species leads to the formation of a hydrogel simply called the "gel." The hydrothermal crystallization of this gel leads to the zeolite being obtained.

The crystallization mechanism is based on the hypothesis of crystal formation in solution. The appearance of germs, then their growth, would be the result of condensation reactions between specified species present in the solution. The renewal of these species would be done by dissolving the solid phase of the gel [14].

2. Simplified description of the ZSM5 structure

Due to the complexity of the elementary lattice of zeolites, the structure of the framework is most often described by an ordered assembly of smaller unitary units [15] called SBU. Only the centers of silicon and aluminum tetrahedra are taken into account in this representation.

In the case of MFI zeolites [11, 16–18], the structure is defined from an arrangement of six SBU 5–1 tetrahedra (**Figure 3a**); SBU5–1 groups combine to form pentasil-like structural units (**Figure 3b**).

The arrangement of these chain groups (**Figure 3c**) leads to the formation of layers of tetrahedrons (**Figure 3d**) generally chosen to schematize the porosity of zeolites ZSM-5.

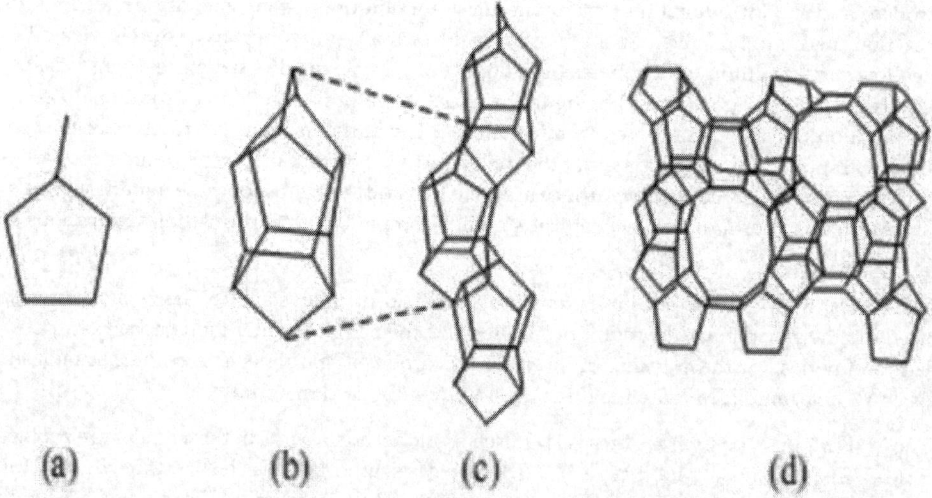

Figure 3. Elements constituting the structure of a ZSM-5 zeolite. (a) Type 5–1 SBU secondary construction unit. Assembly of secondary building units (b) in pentasil group, (c) in chain, and (d) in layers of tetrahedra.

The arrangement of the secondary building units generates a microporous structure composed of straight channels interconnected by sinusoidal channels within the MFI zeolites; **Figure 4** gives an illustration, and the channels have an elliptical opening, whose shape and dimensions vary according to the Si/Al ratio. In the case of zeolite ZSM-5, the right and sinusoidal channels have an opening, respectively, of the order of 0.54×0.56 and 0.51×0.55 nm^2. In the case of silicalite for an Si/Al ratio close to infinity, the sinusoidal channels become almost cylindrical (an aperture diameter of the order of 0.54–0.56 nm) while the right channels conserve an elliptical section with a dimension of the order of 0.51×0.55 nm^2 [19]. Three geometric sites are defined within the framework of MFI zeolites (**Figure 4**):

The sites (I): in the sinusoidal channels, of the order of $0.51 \times 0.55 \times 0.66$ nm^3.

The sites (II): in the right channels with a dimension of $0.54 \times 0.56 \times 0.45$ nm^3.

The sites (III): the intersection of the channels, with a volume of 0.9 nm^3.

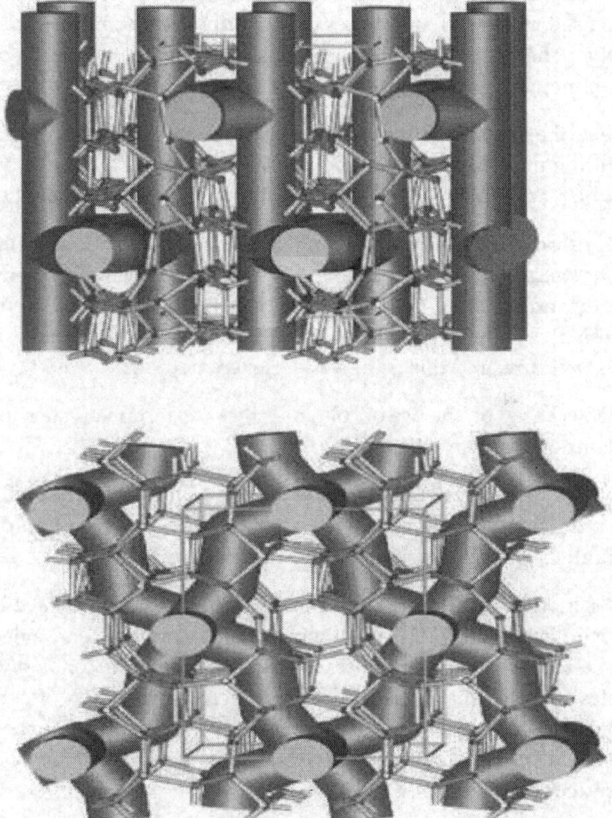

Figure 4. The channels of the ZSM-5.

3. Nanostructures ZSM-5

As already mentioned earlier, microporous materials have been widely used in chemical applications such as catalysis and in separations as adsorbents. However, there is another class which belongs to this category of porous materials, very promising in the catalytic act and in the separation processes; it is well characterized by its large external surface when compared with conventional zeolites, and they are the "nanocrystalline zeolites."

It is well known that the efficiency and selectivity of a porous catalyst depends on their textural and structural characteristics and more precisely on the number of active sites on the external surface in relation to the number of locations accessible via the porous system. Any effect of crystal size of the zeolite is superimposed on the catalytic properties of the material involved, and the mass transfer properties of the catalyst.

For a long time, researchers have been interested in exploring new techniques in catalysis by inserting nanocrystals of zeolites into the framework of catalysts. Such beneficial effects of crystal size can be used in technical processes. So far, only a few examples have been reported in the study on the benefits of zeolite nanocrystals. These are effective even when they are inserted into the framework of the formed catalyst.

These cases relate to the insertion of zeolite beta nanocrystals into hydrocarbon hydrocracking catalysts, resulting in high yield and cracking activity due to the higher outer surface. In many other catalytic reactions, an effect of crystal size of zeolites has been demonstrated.

Aiming at the identification of a suitable route of ZSM-5 zeolite synthesis with acidic properties and crystal diameters of about 100 nm or less, we evaluate several methods of preparation adopted in the study and that been developed by ourselves. Two methods proved to be reproducible and successful. The first involves the hydrothermal crystallization of clear solutions under autogenous pressure, and this is the Van-Grieken method.

The other approach is based on the use of colloidal silicalite-1 grain crystals in open container crystallization (at atmospheric pressure).

The crystal size distribution and the Si/Al content of the products can be well controlled. The products of both types of syntheses could be transferred into their proton forms by conventional means without causing collapse of the crystal structure.

A large number of modifications in the original synthesis procedure have been developed since the discovery of ZSM-5 in 1972 [20]. In most catalytic applications, a decreasing crystal size has a positive effect because it promotes intra-crystalline diffusion. However, the use of very small particles involves the presence of a significant proportion of active sites on the outer surface of the zeolite and which can be detrimental if the effect of shape selectivity is exploited. As a result, a variety of methods have been developed to remove, neutralize, block, or deactivate these acidic sites [21–24].

However, recently, the acidity of the outer surface is considered a property of great interest when the zeolite is intended for the catalysis of reactions that involve bulky molecules (not able to enter the microporous system) such as degradation of polymers or the cracking of heavy oils and in the production of fine chemistry, and so on [25, 26].

This is one of the main reasons explaining the great interest of nanocrystallines with a high external surface developed in recent years [27–34].

The nanocrystallinity in the zeolite domain can be defined as a situation in which the physico-chemical properties of these crystalline materials are largely determined by a greater weight of the atoms in the outer limit of the crystallite. Their difference in properties is no longer negligible [32].

The crystallization process of the nanocrystals of an MFI-type zeolite has been the subject of a number of works in recent years. Most of them focused on the crystallization of silicalite-1 with clear solutions producing crystals with sizes below 100 nm [35–42]. However, the crystallization of a nanocrystalline ZSM-5 in the presence of aluminum sources has not been studied in depth, despite the fact that the presence of aluminum is considered as an essential factor for obtaining materials with a considerable activity in acidic-catalyzed reactions.

Persson et al. [43] have studied the synthesis of crystals of a colloidal ZSM-5 with sizes in the range of 130–230 nm and a narrow particle size distribution. They found that an increase in the concentration of aluminum has caused a decrease in crystal growth, which implies a decrease in the size of the crystals [43].

Van Grieken [44] reported the synthesis of a ZSM-5 zeolite with sizes in the range of 50–100 nm at 170°C and at an autogenous pressure. He concluded that high alkalinity, a significant amount of water, and the presence of alkaline cations (such as Na^+) are detrimental factors in obtaining nanoparticle zeolites. During the first hours of synthesis, the formation of an amorphous gel phase was detected, composed of particles with sizes below 10 nm. These primary units undergo an aggregation process to report secondary particles at sizes around 20 nm. The aggregation of the secondary particles leads to the formation of the final crystals of the zeolite with sizes in the range of 50–100 nm.

Reding [45] has compared different ZSM-5 nanostructure synthesis methods, concluding that the Van Grieken process [44] is completely reproducible and produces a product with good crystalline properties and moderately 90-nm crystal sizes.

4. Synthesis of nanostructure zeolites

Van Grieken [44] and Jacobsen [46] have given necessary descriptions of their methods of synthesis. Similarly, Verduijn's syntheses have been published in several patents. The possibility of synthesizing ZSM-5 with aluminum has been well recommended:

1. Synthesis from Verduijn [47] consists of a synthesis mixture of molar composition (9.12 TPA_2O, 60 SiO_2, 0.5 Al_2O_3, 936 H_2O) prepared by successively adding aluminum sulfate ($Al_2(SO_4)_3 \cdot 18H_2O$, Merck) and silica to the solutions of tetrapropylammonium hydroxide (20% in water, Fluka). After stirring for 10 min under reflux, a homogeneous solution is

obtained. The hydrogel obtained was cooled to room temperature, and the mass loss due to evaporation is compensated by the addition of the deionized water.

The clear solution is poured into flasks equipped with reflux condensers and mounted in oil baths. The crystallization took place under atmospheric pressure and at static conditions for 12 days. The temperature is adjusted to 80°C.

The products are analyzed at different crystallization times. The products are recovered by centrifugation (at 5000 rpm), and the final product is washed several times with deionized water and then dried overnight at 120°C.

The result of this work is reflected in **Figure 5** which shows aggregates of nanoparticles.

2. Synthesis according to Van Grieken [44] consists of using aluminum sulfate ($Al_2(SO_4)_3 \cdot 18H_2O$, Merck), tetrapropylammonium hydroxide (TPAOH, 20% wt in water, Fluka), and tetraethylorthosilicate (TEOS, Sivento) to produce final synthesis solutions of composition: (10.7 $(TPA)_2O$, 60 SiO_2, 0.5 Al_2O_3). The crystallization is stopped after 48 h, and the product recovery is carried out as the above synthesis (**Figures 6** and **7**).

3. Synthesis of Jacobsen [46], the procedure is described using aluminum sulfate ($Al_2(SO_4)_3 \cdot 18H_2O$, Merck), tetrapropylammonium hydroxide (20%, Fluka), tetraethylorthosilicate (Sivento), and Printex L6 which is a carbon black with a pore volume of 1.2 mL/g (Degussa).

   The synthesis mixture is inserted into the pores of the carbon black with the following composition: (10.8 $(TPA_2)O$, 60 SiO2, 0.5 Al_2O_3: 468 H_2O: 240 EtOH). The use of carbon black consists in limiting crystal growth of the formed zeolite.

4. Atmospheric crystallization [48] with silicalite-1 grains involves one atmospheric crystallization procedure with grains of silicalite-1 used as seeds. This method of synthesis comprises two crystallization steps. In the first, a colloidal silicalite-1 is prepared according to the Verduijn procedure as described earlier, without the addition of an aluminum source.

   After crystallization for 72 h at 80°C under atmospheric pressure, the nanoparticles of silicalite-1 were separated by centrifugation and decantation several times. In a second step, new synthesis mixtures were prepared by adding aluminum sulfate and silica to a solution of tetrapropylammonium hydroxide in water (20 wt%, Fluka), after 10 min of stirring under reflux and cooling.

   The solution was seeded with colloidal silicalite-1 previously prepared. The resulting composition of the mixtures synthesis is as follows: (9.12 TPA2O, 60 SiO2, 0.5 Al2O3, 936H2O).

   With: (60 (SiO2) = a (SiO2) silicalite-1 + b (SiO2) solution).

   The following images are taken with MBE scanning electron microscope, and this technique shows the existence of nanocrystals:

   A: with a sowing of 10% of the weight in silicalite-1, and 8 days of crystallization;

   B: with a sowing of 33% of the weight in silicalite-1, and 8 days of crystallization (**Figure 8**).

5. Belarbi et al. [49] tried the syntheses of the nanostructures ZSM5 in alkaline-fluorinated medium, instead of the hyroxyl medium used by the previous researchers.

The alkaline-fluorinated medium produces smaller crystal particles with crystallization faster compared to the first medium. As part of this work, they realized the synthesis of the ZSM-5 nanocrystals with the alkaline cation K^+, and the use of this cation seems to activate the kinetics crystallization of ZSM-5 [50]. The use of fluorinated medium allowed the incorporation of insoluble elements such as Co^{2+}, Fe^{3+}, and Ti^{4+} carried out with moderation in a hydroxyl medium.

Figure 5. SEM image of a nanocrystalline ZSM-5 synthesized according to the Verduijn method [44] with the crystallization time of 12 days.

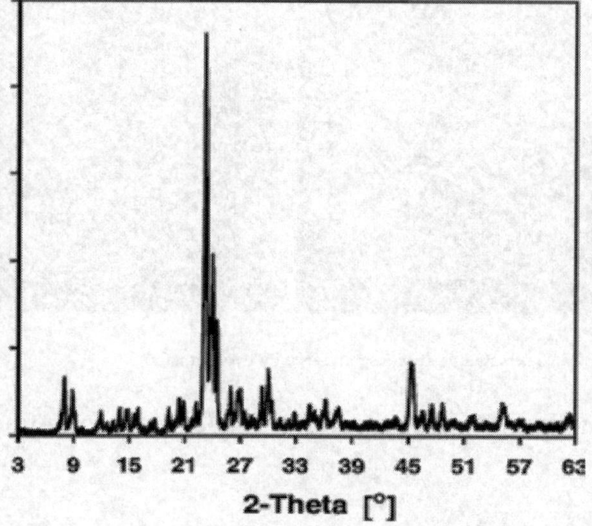

Figure 6. (a) DRX spectrum of a nanocrystalline ZSM-5 according to the Van Grieken method [44] with 12 days as the crystallization time.

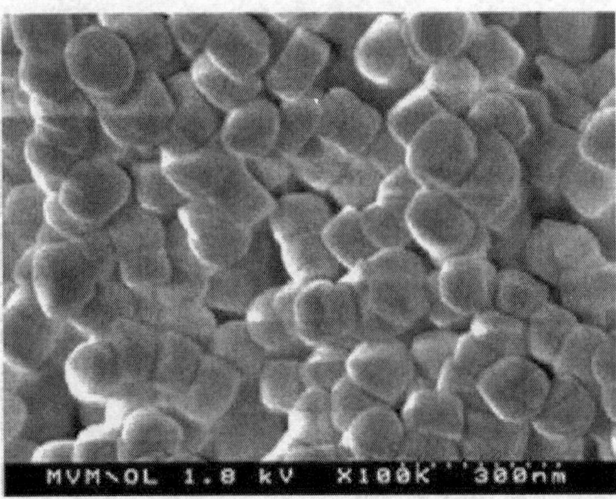

Figure 7. MEB image of a nanocrystalline ZSM-5 synthesized according to the Van Grieken method [44] with 12 days as the crystallization time.

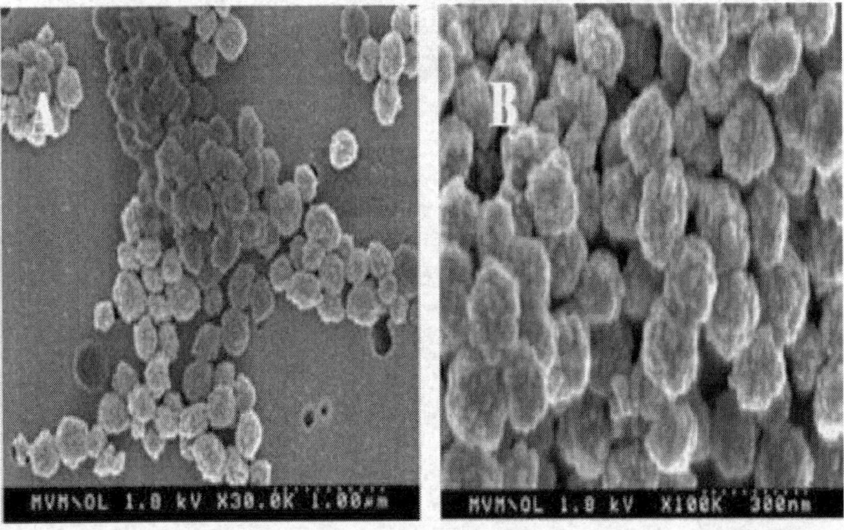

Figure 8. MEB image of a nanocrystalline ZSM-5 synthesized by two methods.

5. Synthesis method of nanostructures ZSM5 in alkaline-fluorinated medium

The following chemical products sodium silicate (63% SiO_2, 18% Na_2O, 18% H2O), aluminum sulfate octadecahydrate ($Al_2(SO_4)3{\cdot}18H_2O$), potassium fluoride (KF), tetrapropylammonium bromide (TPBr), demineralized water, and sulfuric acid H_2SO_4 1 N have been used to obtain ZSM5 nanostructures.

From a given molar composition, the following synthesis protocol has been involved [49]:

1. Solution (1): obtained by dissolving sodium silicate in deionized water under strong agitation at 60°C.
2. Solution (2): aqueous solution consisting of aluminum sulfates and potassium fluoride under intense agitation.
3. Solution (3): obtained by dissolution of tetrapropylammonium bromide (TPABr) in deionized water with stirring at room temperature.

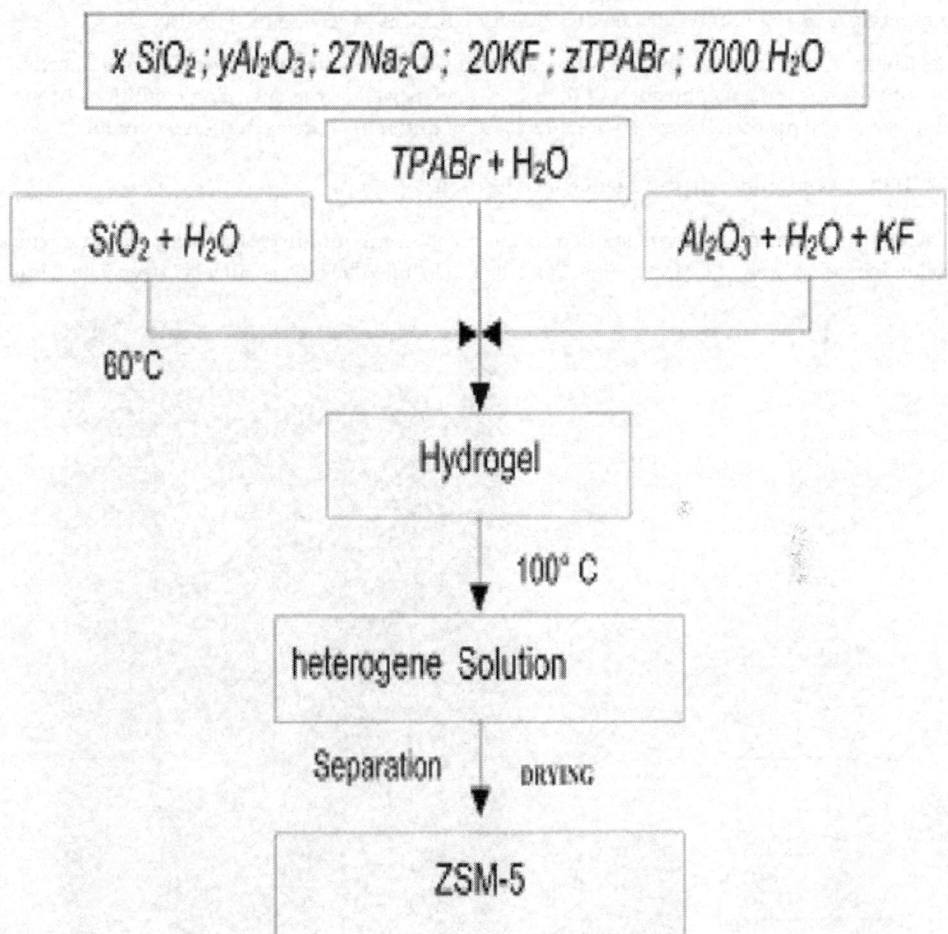

Figure 9. Experimental protocol.

Still stirring, solutions (1) and (2) are poured dropwise on solution (3). The reaction scheme in **Figure 9** summarizes these steps.

The pH of the reaction mixture measured with a pH meter is usually in the order of 12. It is adjusted between 11 and 10.5 by the addition of sulfuric acid 1 N H_2SO_4. The hydrogel obtained is placed in a ground flask, immersed in a crystallizer containing silicone oil (**Figure 10**). The crystallization of the zeolite is carried out under reflux under dynamic conditions (with continuous stirring). The synthesis is carried out at reflux of the reaction mixture (at 100°C), at atmospheric pressure, and for crystallization times ranging from 1 to 6 days.

At the end of crystallization, a biphasic (heterogeneous) solution was obtained, containing an aqueous and a solid phase. The material collected after centrifugation was washed several times with deionized water up to neutral pH. The obtained zeolite was dried overnight at a temperature of 100°C and calcined at 550°C for 8 h, so as to release its porosity.

Belarbi et al. realized the synthesis of nanocrystals ZSM-5 by varying several factors in order to optimize the mole composition of their hydrogel as well as the operating conditions of the syntheses. The products obtained are analyzed by different characterization techniques.

5.1. Influence of mineralizing agent concentration

The content agent mobilizer is studied in the reaction mixture from the composition of the following gel: 200 SiO_2, 1 Al_2O_3, 27Na_2O, x KF, 20 TPABr, 7000 H_2O (with x between 5 and 40).

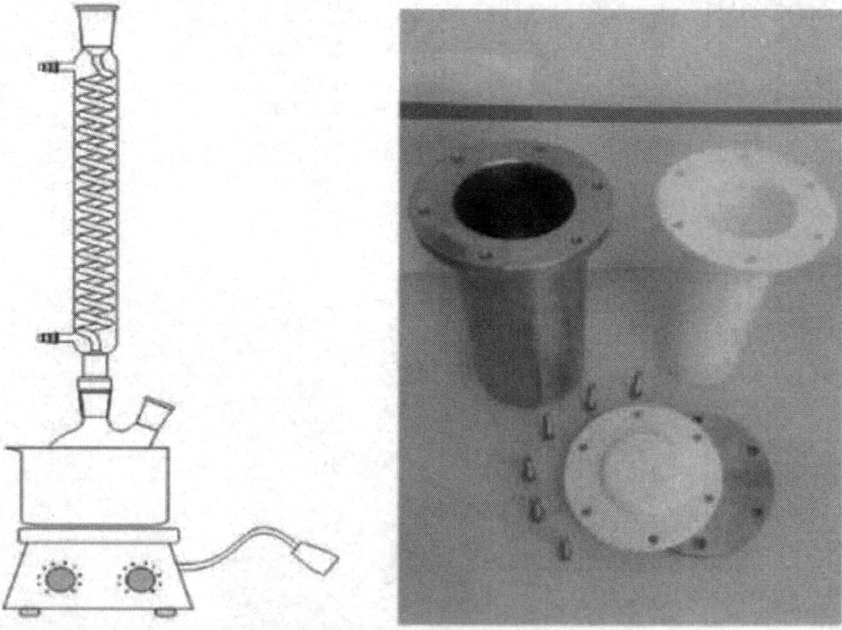

Figure 10. Different synthesis reactors (right autoclave) (left reflux system).

The experimental protocol is described previously; the spectroscopic analysis data by XRD showed that a crystallinity of 100% is recorded for the concentration of 20 moles of KF; beyond this concentration, a significant reduction of the crystallinity rate is observed. The effect of F variation on the crystallinity and quality of the product obtained is proposed by Guth [14], and the increase in the concentration of F ions leads to a growing number of defects in the structure of the product obtained (**Figure 11**).

5.2. Study of the aging time effect

The variation in the time aging was also optimized, and the step precedes the crystallization under reflux with continuous agitation. For a composition of the following reaction mixture: 100 SiO_2/Al_2O_3, 20 KF, 7000 H_2O, 27 Na_2O, and 20 TPABr at 100°C, the duration of crystallization was 8 days, and the aging time was varied from 0 to 92 h (**Table 1**).

The maturing time plays a very important role on the synthesis of zeolites by mainly decreasing the induction period and promoting the formation of the first germs.

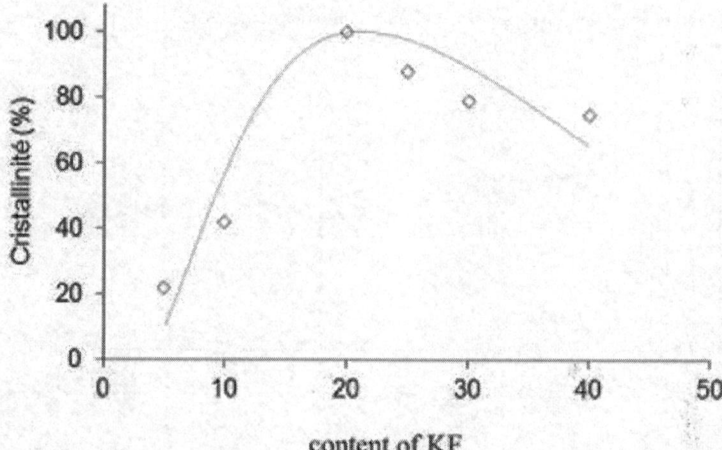

Figure 11. The effect of KF concentration variation on the crystallinity level of ZSM-5.

Samples	Aging time (h)	Crystallinity rate (%)	Diameter (nm)
1NZSM5	0	75	—
2NZSM5	20	95	52*
3NZSM5	44	100	<25*
4NZSM5	68	98	<60*
5NZSM5	62	92	41*

*Estimation using MET.

Table 1. Effect of aging time on structural property.

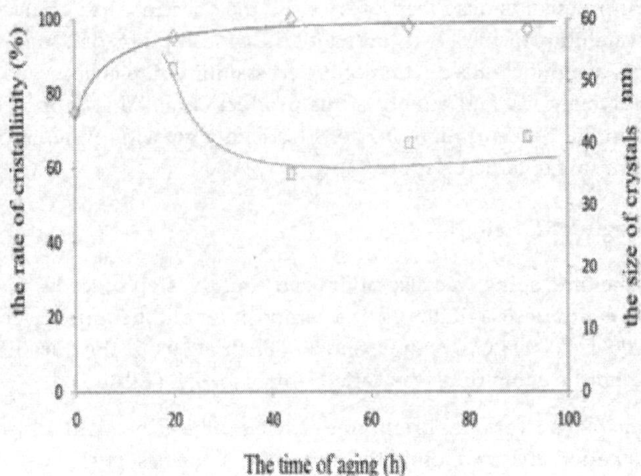

Figure 12. The effect of aging time on crystal size and crystallinity rate of the ZSM5.

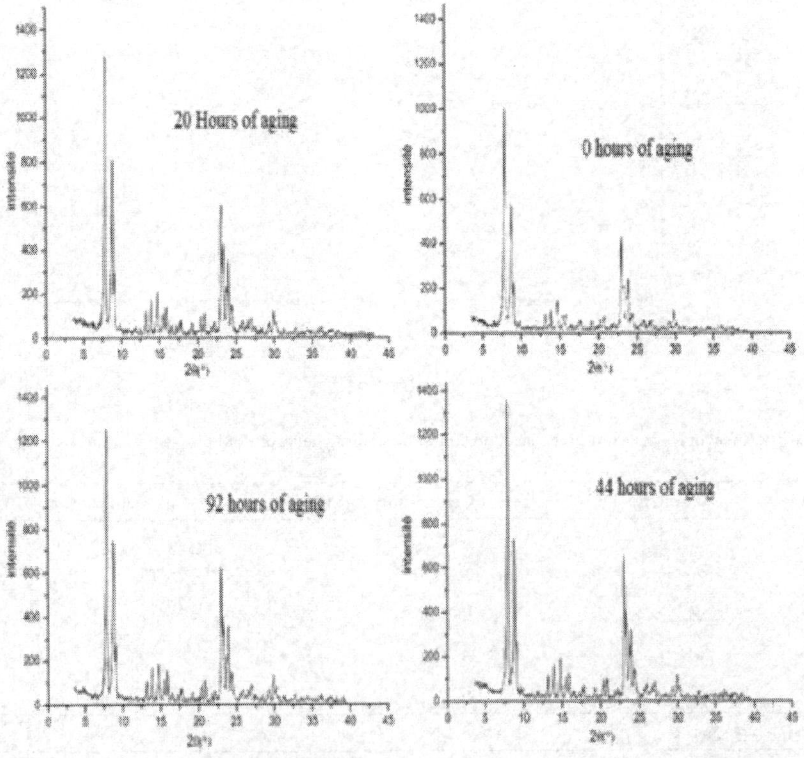

Figure 13. X-ray diffractogram of ZSM-5 at different aging times.

It was found in this study that when the hydrolysis time varied from 0 to 20 h, the crystallinity rate of ZSM-5 increased slightly; for an aging time of 44 h, the crystal size falls down from 52 to less than 25 nm; and for a longer aging period (samples 4 and 5), no appreciable change was observed. The supersaturation of the gel reaches its maximum after only 2 days of aging (**Figures 12** and **13**).

To better determine the size and morphology of the crystals, the transmission electron microscopy (TEM) technique was used; according to the TEM images represented in **Figure 14**, the prepared ZSM-5 samples consist of very small crystals, with size that varies from 15 and 25 nm. These crystals are not present as isolated particles, but in the form of small aggregates with sizes that exceed 200 nm.

5.3. The optimization of the organic template quantity

The quantity of organic template was optimized by varying its concentration from 0 to 40 moles, while keeping the ratio SiO_2/Al_2O_3 = 100, KF = 20 moles, H_2O = 7000 moles, Na_2O = 27 moles at 100°C for a period of crystallization of 6 days (**Table 2**).

The choice of TPABr (tetrapropylammonium bromide) as a structuring agent in the synthesis of zeolite ZSM-5 was been widely cited in the study [51]. This is due to the fact that this template plays a real role of template; in addition, it directs toward the structure ZSM-5. It

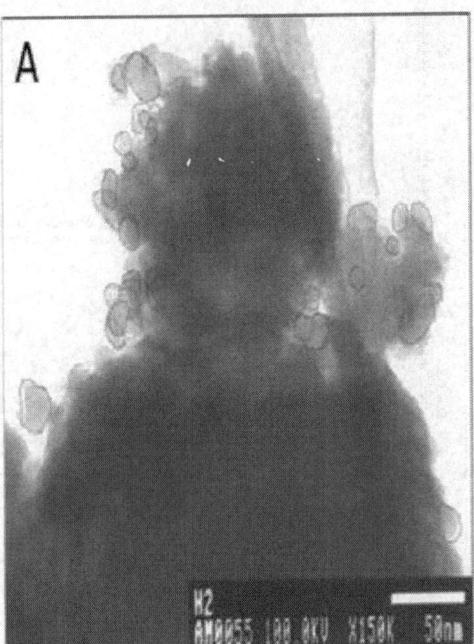

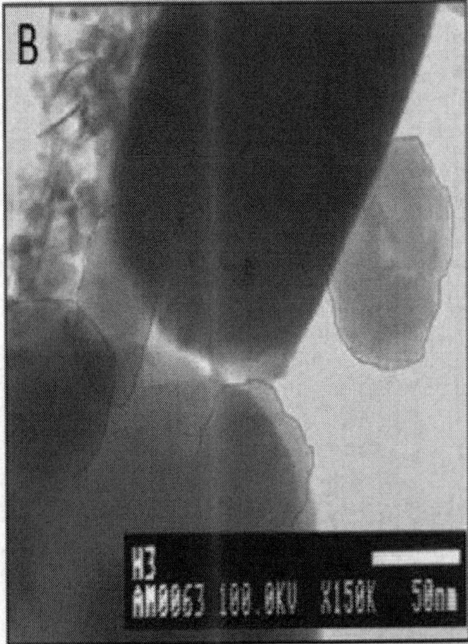

Figure 14. TEM (transmission electron microscopic image) of (A) N3ZSM5 with an aging time of 44 h (2 days) and (B) N4ZSM5 with an aging time of 68 h (3 days).

participates in the assembly SBU_{5-1} units and subsequently contributes to their stability. On the other hand, the geometry of the TPABr molecule promotes the rapid crystallization of ZSM-5 compared to the other organic structuring agents. This is explained by the low activation

Samples	$TPABr/SiO_2$	Rate of crystallinity (%)	Diameter (nm)
N6ZSM5	0.4	100	20
N7ZSM5	0.3	82	33
N8ZSM5	0.25	78	48
N9ZSM5	0.2	71	52
N10ZSM5	0.1	30	70
N11ZSM5	0.05	21	—
N12ZSM5	0	5	—

Table 2. Effect of the template agent content on the size of formed crystals.

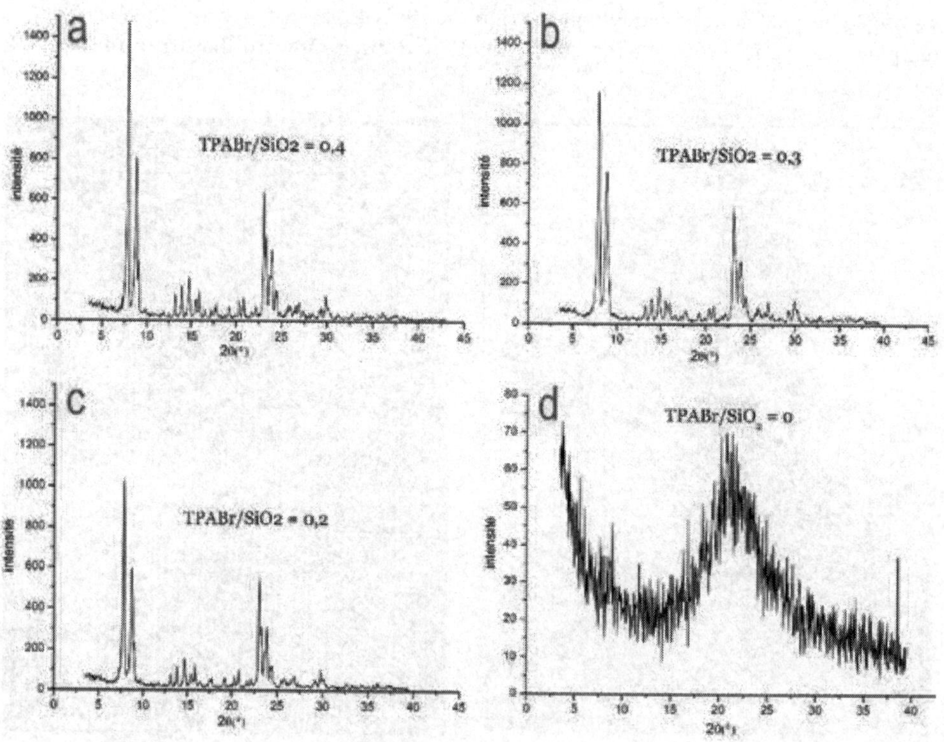

Figure 15. X-ray diffractograms at different TPABr/SiO2 ratios.

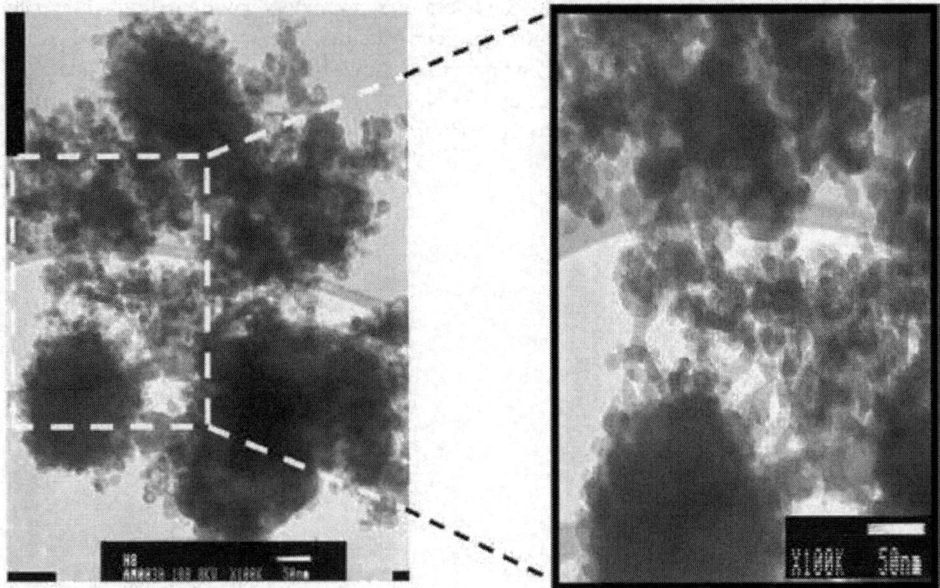

Figure 16. TEM images of N6ZSM5 sample with $TPABr/SiO_2$ ratio = 0.4.

energy of this template E = 83 kJ/mole [52], compared, for example, with that of TEABr (tetraethylammonium bromide), which is equal to 90 kJ/mole [53].

Figure 15(a–d) represent the XRD spectra of the samples synthesized with TPABr concentrations ranging from 0 to 40 moles. The analysis of these spectra shows that the envelope of peaks between 7 and 10° (2θ) and between 22 and 25° increases in intensity as the concentration of TPABr increases in the reaction mixture.

Sample A (**Figure 16**) was placed under the probe of an electron microscope transmission with sufficient magnification to see and even determine the size of the crystals which touches the field of nanometers, some of these crystals browse the size of 10 nm.

6. Conclusions

The screening of the research works concerns the synthesis of zeolitic nanostructures of the ZSM5 type, shown that in spite of the chemical elements that constitute these products which are the same, namely SiO_2, Al_2O_3, TPA_2O, H_2O, and Na_2O; however, the changes carried by the different authors in the source of these elements, the experimental protocol, and the methods of preparation allow to considerably affect the final size of the crystals of these zeolites. The work carried out for the synthesis of ZSM5 nanostructures in an alkaline fluoride medium has shown that certain operating parameters have a great influence on the quality of the final product and on the size of these crystals. The operating parameters having a direct influence on the synthesis which include the mobilizing agent content F, the amount of optimal structuring agent used, and the aging time. The final etude shows that the optimum concentration is 20 KF, 40 TPABr, 100 SiO2, 7000 H2O, and 27 Na2O. For an aging time of 44 h and for a crystallization time of 6 days and a crystallization temperature of 100°C, the crystal size of the nanostructures ZSM5 obtained was reduced until 25 nm.

References

[1] Cronstedt A. Kongl.Svenska Vetenskaps Academiens Handlingar. Vol. 17. Stockholm: Lorentz Ludvig Grefing; 1756. p. 120

[2] Nagy JB, Bodart P, Hannus I, Kiricsi I. Synthesis, Caraterization and Use of Zeolitic Micoporous Mateials. Hungary: DecaGen Ltd; 1998

[3] Breck DW. Zeolite Molecular Sieves Structure Chemistry and Use. New York: John Wilay & Sons; 1974

[4] Baerlacher C, Meier RW, Olson DH. Atlas of Zeolite Structure Types. 3rd Revised ed. London: Butterworth-Heinemann; 1992

[5] Barrer RM, Baynham JW, Bultitude FW, Meier WM. Hydrothermal chemistry of the silicates. Part VIII. Low-temperature crystal growth of aluminosilicates, and of some gallium and germanium analogues. Journal of the Chemical Society. 1959:195

[6] Lopez A. thèse de doctorat, université de haut alsace; 1990

[7] Milton RM. Rapport in US patent 2; 1959

[8] International Zeolite Association in http://www.iza-structure.org/databases/

[9] Meier WM, Olson DH. Atlas of Zeolite Structure Types. London: Butterworth; 1987

[10] Sie ST. Past, present and future role of microporous catalysts in the petroleum industry. Studies in Surface Science and Catalysis. 1994;**85**:587

[11] Kokotailo GT, Lawton SL, Olson DH, Meier WM. Structure of synthetic zeolite ZSM-5. Nature. 1978;**272**:437

[12] Wu EL, Lawton SL, Oison DH, Rohrman AC, Kokotailo GT. ZSM-5 type materials. Factors affecting crystal symmetry. The Journal of Physical Chemistry. 1979;**83**:2777

[13] Argauer RJ, Kensington M, Landolt GR, Audubon NJ. Crystalline Zeolite ZSM-5 and Method of Preparing the Same. Rapport in US patent Office; 1972

[14] Guth JL, Gaullet H. Journal de Chimies Physique. 1986;**83**:155

[15] Meier WM. Zeolite structure. Journal of the Society of Chemical Industry, London. 1968

[16] Barrer RM. Chemical nomenclature and formulation of compositions of synthetic and natural zeolites. Pure and Applied Chemistry. 1979;**51**:1091

[17] International Zeolithe Association. Database of Zeolite Structures. http://www.iza-structure.org

[18] Olson DH, Kokotailo GT, Lawton SL. Crystal structure and structure related properties of ZSM-5. The Journal of Physical Chemistry. 1981;**85**:2238

[19] Ashtekar S, McLeod AS, Mantle MD, Barrie PJ, Gladden LF, Hastings JJ. The Journal of Physical Chemistry. B. 2000;**104**:528

[20] Argauer RJ, Landolt GR. US Patent 3702886, Mobil Co. 1972

[21] Uguina MA, Sotelo JL, Serrano DP, Van Grieken R. Industrial and Engineering Chemistry Research. 1992;**31**:1875-1880

[22] Uguina MA, Sotelo JL, Serrano DP. Canadian Journal of Chemical Engineering. 1993;**71**: 558-563

[23] Roger HP, Kramer M, Moller KP, Connor CTO. Microporous and Mesoporous Materials. 1998;**21**:607-614

[24] Weber RW, Moller KP, Connor CTO. Microporous and Mesoporous Materials. 2000;**35-36**: 533-543

[25] Aguado J, Serrano DP, Sotelo JL, Van Grieken R, Escola JM. Industrial and Engineering Chemistry Research. 2001;**40**:5696-5704

[26] Corma A. Journal of Catalysis. 2003;**216**:298-312

[27] Madsen C, Jacobsen CJH. Chemical Communications. 1999:673-674

[28] Schmidt I, Madsen C, Jacobsen CJH. Inorganic Chemistry. 2000;**39**(11):2279-2283

[29] Rakoczy RA, Traa Y. Microporous and Mesoporous Materials. 2003;**60**:69-78

[30] Holmberg BA, Wang H, Norbeck JM, Yan Y. Microporous and Mesoporous Materials. 2003;**59**:13-28

[31] Li Q, Creaser D, Sterte J. Chemistry of Materials. 2002;**14**(3):1319-1324

[32] Camblor MA, Corma A, Valencia S. Microporous and Mesoporous Materials. 1998;**25**: 59-74

[33] Corma A, Fornes V, Perguer SB, Maesen TLM, Buglass JG. Nature. 1998;**396**:353-356

[34] Corma A, Fornes V, Martınez Triguero J, Pergher SB. Journal of Catalysis. 1999;**186**:57-63

[35] Persson AE, Shoeman BJ, Sterte J, Otterstedt J-E. Zeolites14. 1994:557-567

[36] Schoeman BJ, Sterte J, Otterstedt J-E. Zeolites. 1994;**14**:568-575

[37] Burkett SL, Davis ME. The Journal of Physical Chemistry B. 1994;**98**:4647-4653

[38] Tsuji K, Davis ME. Microporous Materials. 1997;**11**:53-67

[39] Schoeman BJ, Regev O. Zeolites. 1996;**17**:447

[40] Schoeman BJ. Microporous Materials. 1997;**9**:267

[41] Li Q, Mihailova B, Creaser D, Sterte J. Microporous and Mesoporous Materials. 2000;**40**: 53-62

[42] Li Q, Mihailova B, Creaser D, Sterte J. Microporous and Mesoporous Materials. 2001;**43**: 51-59

[43] Persson AE, Schoeman BJ, Sterte J, Otterstedt JE. Microporous and Mesoporous Materials. 1995;**15**(7):611-619

[44] Van Grieken R, Sotelo JL, Menendez JM, Melero JA. Microporous and Mesoporous Materials. 2000;**39**:135-147

[45] Reding G, Maurer T, Kraushaar-Czarnetzki B. Microporous and Mesoporous Materials. 2003;**57**:83-92

[46] Jacobsen CJH, Madsen C, Janssens TVW, Jakobsen HJ, Skibsted J. Microporous and Mesoporous Materials. 2000;**39**:393

[47] Verduijn JP. WO 97/03019, to Exxon Chemical Patents Inc. 1997

[48] Tosheva L, Valtchev VP. Nanozeolites: Synthesis, crystallization mechanism, and applications. Chemistry of Materials. 2005;**17**:2494-2513

[49] Belarbi H, Lounis Z, Hamacha R, Bengueddach A, Trens P. Colloids and Surfaces A: Physicochemical and Engineering Aspects. 2014;**453**:86-93

[50] Van Koningsveld H, Van Bekkum H, Jansen JC. Zeolite. 1990;**10**:235-245

[51] Mostowicz R, Sand LB. Zeolites. 1982;**2**:143

[52] Patarin J, Kessler H, Guth JL. Zeolites. 1990;**10**:674

[53] Corma A, Rey F, Rius J, Sabater MJ, Valencia S. Nature. 2004;**431**:287

Section 2

Properties and Applications of Nanocrystals

Chapter 4

Colloidal Solutions with Silicon Nanocrystals: Structural and Optical Properties

Abstract

In this work, colloidal solutions with silicon nanoparticles using different solvents were synthetized. Structural, morphological and optical characterizations were realized, and these were studied. X-ray diffraction (XRD) was used to measure the diffractograms of the colloidal solutions, which are composed of silicon nanocrystals (Si-ncs), with an average size of approximately 3 nm, and a preferential crystalline orientation (311). Atomic force microscopy (AFM) images show that the morphology of silicon nanoparticles (Si-nps) is agglomerated in a big amount, which is corroborated by means of the roughness. On the other hand, high resolution transmission electronic microscopy (HRTEM) images show on average size of the Si-nc ranging from 1.5 to 10 nm, which depends on the solvent used. Also, different preferential crystalline orientations of the Si-nc such as (311), (220) and (111) were obtained. A correlation between the optical and structural properties was realized in colloidal solutions with silicon nanoparticles and different solvents.

Keywords: silicon nanocrystals, colloidal solutions, porous silicon, XRD, AFM, HRTEM

1. Introduction

In the last years, the synthesis of the silicon nanostructures has been developed in grand manner, due to both by their interesting quantum effects that these structures present and also by their interesting optical and unique electrical properties. A fundamental result that has been found in the silicon nanocrystals (Si-ncs) is that they make a contribution to the shifting in the prohibited energy band. Thus, we observed in the measurements that the energy band gap increases when decreasing the Si-nc size [1–3]. Other interesting result is related to the emission efficiency of light in the visible range, which is linked to the energy band gap of the Si-nc, which in turn depends strongly on the size of the Si-ncs, and when they have lesser size than the Bohr exciton radius in the silicon, the spatial confining of the carriers in the nanocrystal is greater, which gives rise to a strong overlap of the wave functions in the k-space in both electrons and holes. Therefore, the nanocrystals can absorb and emit light with different energies by controlling only their size [4]. Silicon nanocrystals have been widely exploited in electronics and other areas, due to their low toxicity, and they exhibit the ability to be doped in order to become either an n-type or p-type material [5]; this fact allows generating new technologies such as sensors [6–8], biosensors [9–12], magnetic materials [13], photodiodes [14], Bragg reflectors [15], photonic applications [16–21], nonvolatile storage devices [22–24], solar cells of third generation [25–29] as well as tandem solar cells [30–35], among other devices [35, 36].

Actually, it exists a big interest in the topic of photovoltaic energy, due to Queisser-Shockley theory limit, in one standard solar cell of Si where there is a loss of efficiency due to the phonon scattering phenomenon that is produced by the hot electrons. To eliminate such undesirable effects, it has been proposed two alternatives for improving the efficiency of such Si cell. The first one consists of making heterostructures with materials of different energy band gaps; in each layer forming the heterostructure, the energy will be absorbed and transformed into electric energy. The second one is based on the down conversion process that consists of depositing one layer on the solar cell and using the photoluminescent effect to reduce the energy of incident photons. These photons make a more efficient conversion of the solar energy into electric energy. These two ways proposed require a careful control of the size of the nanocrystals [4].

Today, there are a lot of methods for obtaining silicon nanocrystals, namely Stober, pulsed laser, controlled precipitation, emulsions, oxidation, silane combustion, gas evaporation, co-sputtering and thermic degradation [37–48], chemical vapor deposition (CVD) [49, 50], low pressure chemical vapor deposition (LPCVD) [51], ionic implantation [52, 53] and other ones. Unfortunately, such methods are very expensive. As an alternative, a method consisting of using colloidal solutions for obtaining Si-ncs may be used. We have reported somewhere previously how to obtain these colloidal solutions [54]. In this chapter, we report the observed structural and morphological properties of silicon nanocrystals in different organic solvents as colloidal solutions with Si-ncs, and we quantify the exact size of these nanocrystals.

2. Experimental method

The first step to obtain the silicon nanocrystals consisted on obtaining the porous silicon (PSi). To synthetize the PSi, we used the electrochemical etching process for which was used hydrofluoric acid (HF) with different electrical current densities. The obtained PSi samples were subjected to a scrapped-off process, and after this, a grinding process was carried out in an agate mortar.

After the latter process a fine powder was obtained, it was mixed with organic solvents (acetone, methanol and ethanol) and then a process of decanting over the larger particles found in the solution was done. Finally, almost transparent solutions were obtained as observed at a glance, but when they were exposed under UV radiation, they showed an efficient photoluminescence lying between green and blue colors. The obtaining of nanocrystals by means of colloidal solutions was previously reported by the authors [54]. Therein, the process used to obtain the nanocrystals is explained in detail.

The structural properties of the silicon nanocrystals were determined by the high resolution transmission electronic microscopy (HRTEM) technique with a JEOL-2010 HRTEM (Jeol, Tokyo, Japan) system with a potential of 200 kV. The X-ray diffractograms were obtained with Discover Bruker D8 equipment operated at 40 kV and at 40 mA using a CuKa radiation (1.5406°), with a Lynxeye detector and a 0.2 mm divergence grid as primary optics and a grid of 3 mm anti-dispersion as secondary optics. The morphological properties were studied by atomic force microscopy (AFM) with a JEOL JSPM-5200 team.

3. Results and discussion

3.1. X-ray diffraction (XRD)

To obtain the diffractograms of the colloidal solutions, we deposited a colloidal solution on silicon substrates with crystalline orientation (100), so that they functioned as support to the Si-ncs. The diffraction pattern peaks obtained in the diffractogram are localized near those that have been reported in Ref. [55]; the case of crystallographic planes is also similar [56–59]. This result indicates clearly that there are silicon nanocrystals in our silicon substrate. On the other hand, the XRD data show that Si-ncs have a preferential crystalline orientation (311). However, in some samples (M_09 and M_15), additional peaks were found around $2\theta = 28.52°$ (111), 47.31° (220), and this may be due to the existence of some type of metastable state of the silicon nanocrystals on the silicon substrate as indicated in **Figure 1**. By using the average width of the diffraction peak (FWHM), the average size of the nanocrystals could be calculated by applying the Scherrer equation [60, 61]. **Table 1** shows a comparison between the diffracted peaks of JCPDS card 27–1402, with respect to the diffracted peaks belonging to our samples obtained.

Some papers report that these diffraction peaks are originated by the core/shell structures, where such structures are formed from the combination of silicon nanocrystals coated by an oxide layer [57], due to the etching process in which HF was utilized to obtain the PSi and the subsequent atmospheric exposure.

It was observed that the average size of the nanocrystals varied from 2.88 to 3.67 nm according to the data reported in **Table 1**. In the work that was reported in Ref. [54], it is possible to observe that the nanocrystals have the tendency to orientate themselves in the crystallographic plane (111), but with nanocrystals having a larger size. However, when the nanocrystals were directly measured, it was found that their preferential orientation changed to (311) with smaller sizes. This can be attributed to the fact that the sizes of larger crystals

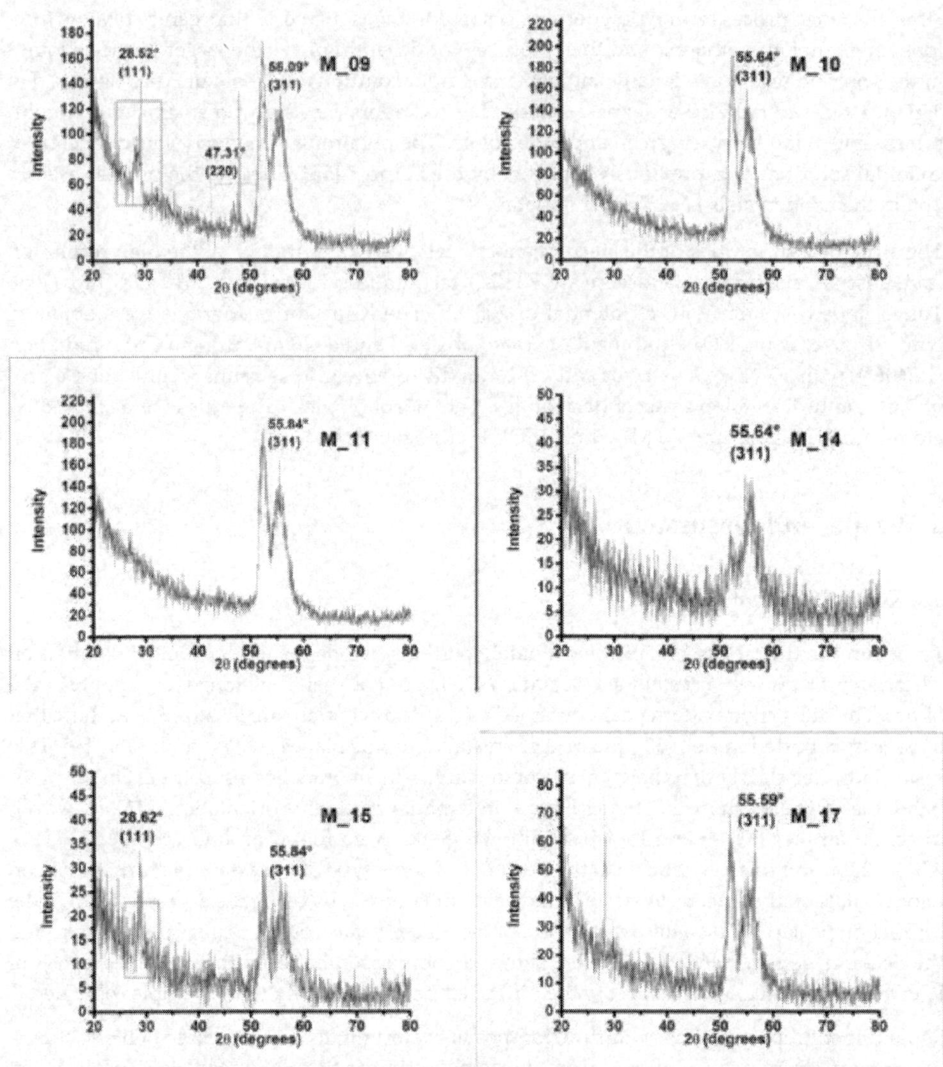

Figure 1. XRD diffractograms of the colloidal solutions obtained.

are due to the agglomerations of small crystals, thus producing a preferential orientation as a consequence of this agglomeration. When we carry out the measurement of the smallest crystal dimensions, in a certain way we measure the crystal dimensions individually; therefore, we can say that we are measuring the true natural orientation of the nanocrystals which is possibly because of the way of preparation of the colloidal solution samples, contrary to other works where they have different orientation tendencies of the crystals such as (111)

Ref. [57]	Sample M_09		Sample M_10		Sample M_11		Sample M_14		Sample M_15		Sample M_19	
Peak position (PP)	PP	Size (nm)	PP	Size (nm)	PP	Size (nm)	PP	Size (nm)	PP	Size (nm)	PP	Size (nm)
28.44	28.52	4.88	N/A	N/A	N/A	N/A	N/A	N/A	28.62	2.19	N/A	N/A
47.30	47.31	4.19	N/A	N/A	N/A	N/A	N/A	N/A	N/A	N/A	N/A	N/A
56.12	56.09	2.88	55.64	3.18	55.84	3.24	55.74	3.06	55.84	3.67	55.59	3.62

Table 1. XRD peaks position and nanocrystal size obtained from the samples.

[60] or an orientation (220) [4]. An important feature which is noteworthy to emphasize is the intensity of the diffraction peaks; such intensity is attributed to the thickness formed by the silicon nanocrystals on the substrate [4]. Therefore, we suggest that the intensity of the peaks depends strongly on the amount of nanocrystals found.

3.2. AFM characterization

The characterization by atomic force microscopy was realized in all colloidal solutions with Si-ncs, in order both to know the roughness of the samples and to demonstrate the existence of silicon nanocrystals by using another characterization technique. The colloidal solution samples were deposited on p-type silicon substrates with orientation (100). The results were the following: all the samples presented particles with spherical shapes well-defined and sometimes irregular shapes, similar to particle agglomerations. The sample M_09 exhibited a set of particles forming several geometries, as can be seen in the XY scale of the image. In this scale, the dimensions of such particles lie in the micron order, but in the Z scale, we find that they have a maximum of 186 nm. Therefore, we can say that there are agglomerates of crystals in the XY plane with micrometric dimensions which have along the Z direction nanometric dimensions. The largest particles have a height of 186 nm and it decreases until finding particles with a height approximately of 18 nm. The 3D image shows the roughness surface of the substrate due to the layer of the colloidal solution that was deposited as shown in **Figure 2**.

By taking a specific area of the image, it was possible to locate an area where particles with heights approximately from 1 to 2.3 nm were found. On the other hand, nanometric particles could also be found in the X-Y plane. Besides, there are agglomerated particles whose size is around 100 nm in the XY plane with a height of approximately 30 nm. This dispersion in size that occurs between the images is due to the sections that are reviewed in the sample as is shown in **Figures 2** and **3**. However, with this information obtained from these images, we can determine and verify that we really have particles of nanometric size.

Similar situation happened for samples M_10, M_11, M_14, M_15 and M_17 because they also had particles with different geometric shapes which in general exhibit a tendency to be spherical. At first glance, it was observed in a large area comprising micron-sized dimensions, but

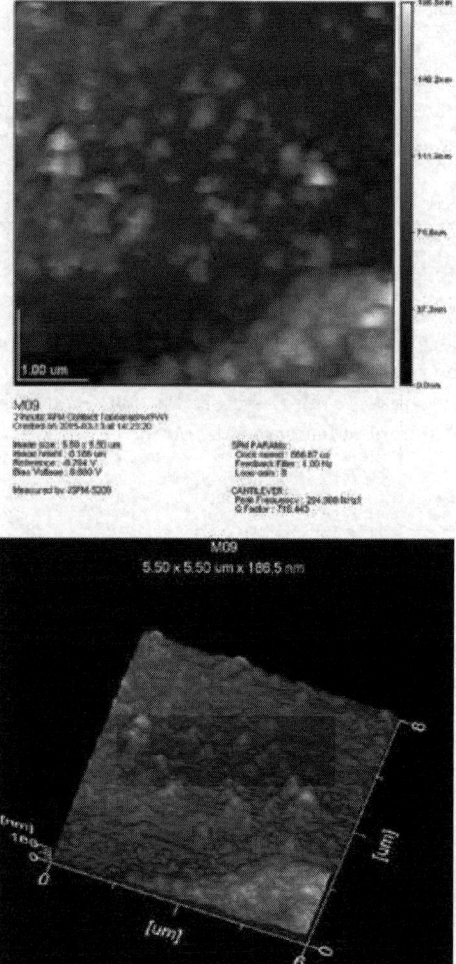

Figure 2. AFM image and 3D image obtained from the sample M_09.

it was also possible to see agglomerated particles with dimensions in the order of hundreds of nanometers, with similar heights. This occurred for some areas studied as shown in **Figure 4**.

On the other hand, by observing specific areas of the samples, particles of smaller sizes could be detected. We could also reduce the scanning area, and we could find a quantity of nanometric particles in the X-Y-Z space with a size of approximately 1.8 nm. These images clearly demonstrate the existence of silicon nanoparticles.

Figure 5 shows the contrast between individual particles that can be found and agglomerates that are produced by the junction of individual particles. The origin of the agglomeration process is not yet clear; we assume that this mechanism may be due to several factors, such as the preparation of the samples, the deposition of the colloids on the substrate or the density of suspended particles in the solvents.

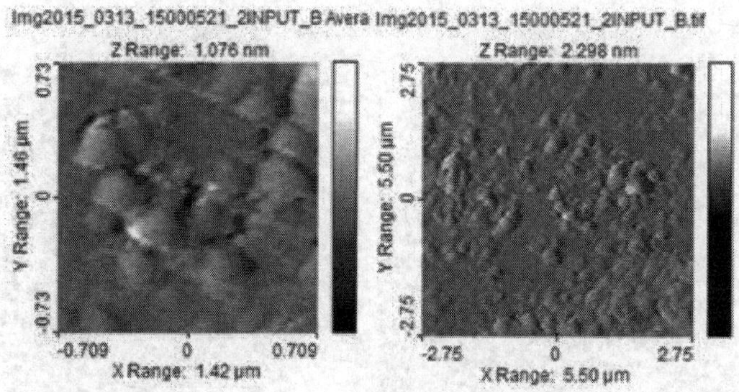

Figure 3. The X-Y plane of the sample M_09 with agglomerations of nanometric size.

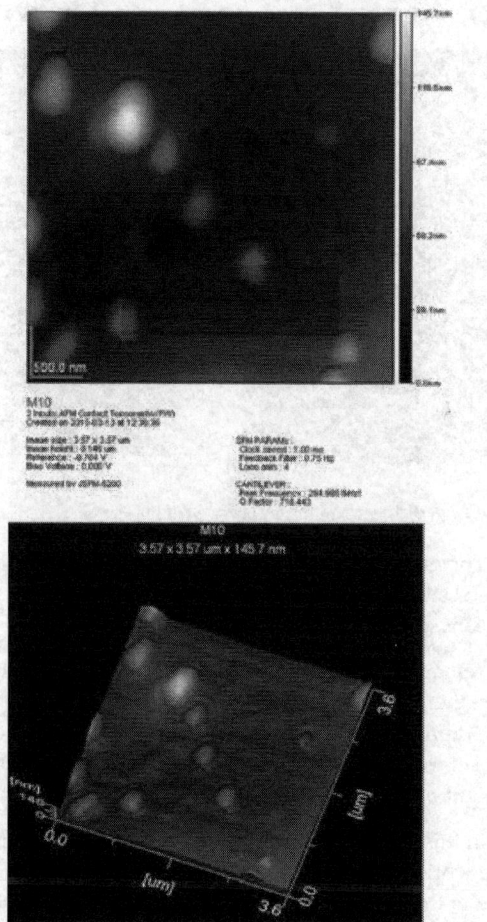

Figure 4. The X-Y plane and 3D image of the colloidal solution sample with Si-ncs.

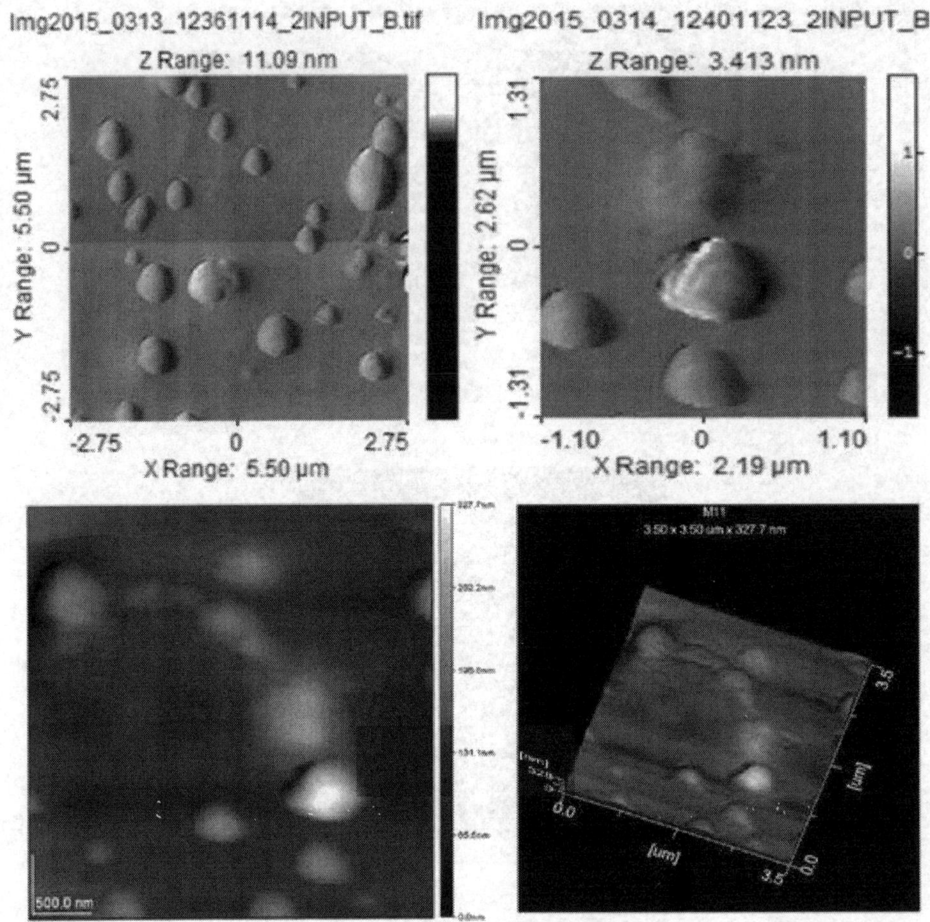

Figure 5. Junction of individual particles generating an agglomeration process.

It is clear that we can corroborate the X-ray diffraction data with these results because in XRD there are average crystal sizes between 1.5 and 3 nm, and it is clear that we are obtaining similar crystal sizes by means of this characterization technique.

For sample M_14, a remarkable fact happened, it consisted of agglomerations of particles, but in a particle image there were longitudinal agglomerations, similar grooves, and on these grooves more particles appeared. The approximate height of these grooves is 100 nm. By closing up these grooves, some particles of the order of 10 nm were found in the X-Y plane having an approximate Z-height of 15 nm (**Figure 6**).

The results obtained in this work are similar to those that have been reported in the literature [61–64]. Most of these works report morphologies similar to those found in this work, including those we reported in the sample M_14, with the formation of grooves, that was similar to that reported by Jasmin et al. [64], because in an image of phase shows a similar behavior, due

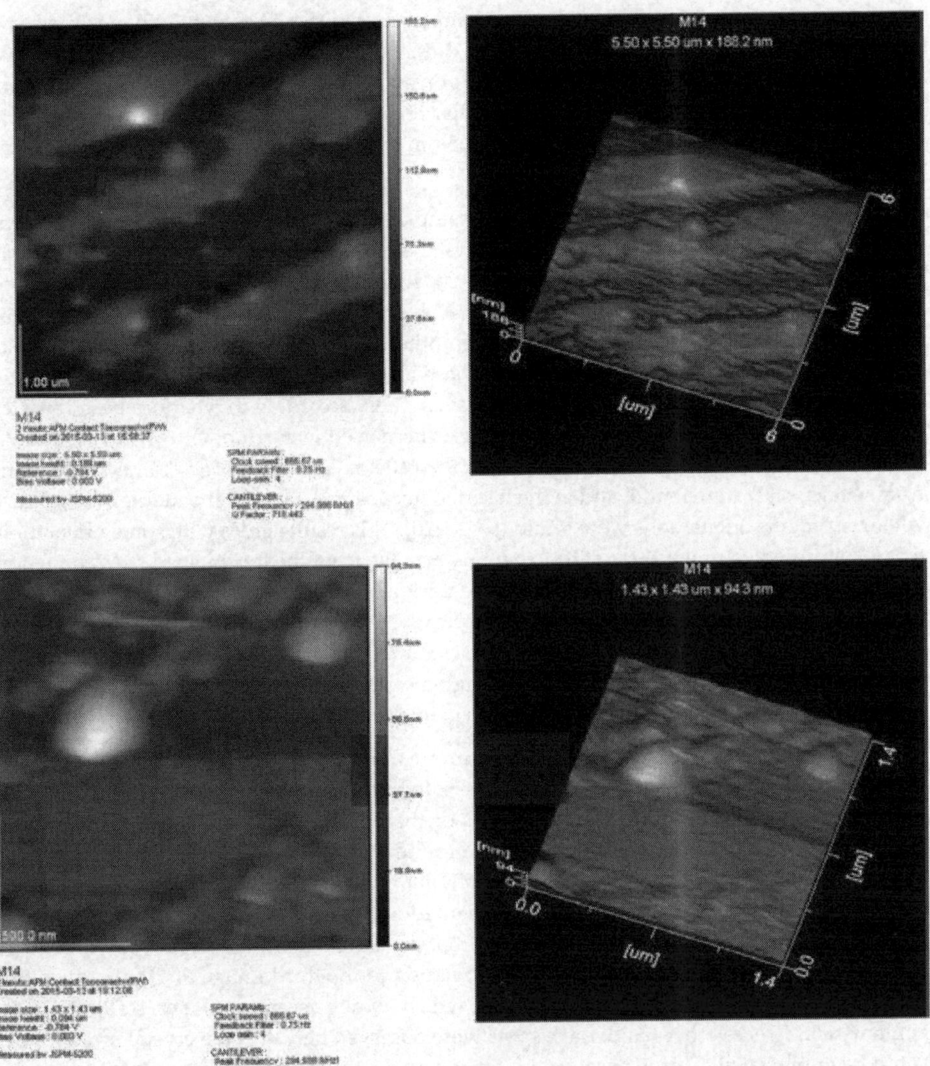

Figure 6. Longitudinal agglomerations, similar grooves, and on these grooves more particles appeared.

to an interference between the sample and the cantilever. However, this behavior is due to the magnetic properties of the particles. Now, in our case, we do not have that kind of properties so we can attribute this behavior to a kind of longitudinal agglomerations of the particles.

3.3. TEM characterization

The results of the HRTEM images of the colloidal solutions show the size, shape and preferential orientation of the silicon nanocrystals immersed in the colloidal solution. This characterization is presented only for three samples with different solvent (ethanol, methanol and acetone). We could note nanocrystals of different diameter sizes in a range from 1.5 to 8 nm and different geometrical shapes; some of which were circular, ellipsoidal and some others with not well defined shapes. Ray Mallar et al. [57] also reported nanocrystals of size close to those obtained in the present work (1.5 nm), with shapes and crystalline orientations similar to the present work.

The three images in **Figure 7** correspond to a sample of colloidal solution with acetone solvent; in two of such images, we can notice a large number of spherical and ellipsoidal nanocrystals and other ones similar to agglomerations (irregular shapes); a few more are not clearly perceived because the scale of the image is very large. Possibly the lack of magnification in this image hid the clearer vision of the nanocrystals, although other works have shown areas with spots or black marks as nanocrystals formation [64]. These black marks correspond to the fact that the nanocrystals have been wrapped in an oxide layer. Then, according to what has been reported in the scientific literature, our sample of acetone is immersed with nanometric nanocrystals. We stress that nanocrystals with dimensions larger than 10 nm were also found. In the image, 6 and 7 nm nanocrystals were found, and in the inserted images, the diffraction patterns are indicated where crystalline orientations were found to be (220). This result agrees with some orientations that were presented in the X-ray diffractograms. Finally, from the third image, we can identify a black spot; the latter is originated because the thickness of that agglomerate impedes the electrons from penetrating into the sample. Consequently, they are bounced in such a region, forming an image similar to an scanning electron microscopy (SEM) one. However, smaller particles were found, which were detached from this agglomerate; this event is consistent with the AFM images, where there are nanocrystals forming agglomerate (**Figure 7**) [54].

The second image of **Figure 8** corresponds to a colloidal solution with ethanol solvent. It could be noted that only a few dark spots were found. This may be possible because either there exists a smaller amount of nanocrystals or there are a lot of nanocrystals but without the oxidant coating. These nanocrystals were found having an average size between 4 and 5 nm, exhibiting circular shapes and other irregular ones. The inserted images of the diffraction patterns depict preferential crystalline orientations corresponding to (220) and (311). This fact confirms us and at the same time proves that the crystal orientations in X-ray diffraction are similar to those obtained by HRTEM. In relation to the third image of **Figure 8**, we find that there is agglomerate, but a little thinner, which allows observing the crystallinity of the agglomerate. On the other hand, dark spots were observed here, but the crystal arrangement could be appreciated. This is because the oxide layer is thinner than the crystal. This proves what was mentioned in the acetone images. It leads one to think that core-shell crystals are formed, where the silicon crystals play the role as a core and the oxidizing layer functions as a shell.

The last images corresponding to **Figure 9** are attributed to a colloidal solution with methanol solvent, where no dark spots were found. In this case, the resolution scale was far better. It allowed that in this image we could find a larger quantity of nanocrystals with a much smaller size than those found in the previous images, showing sizes from 1.4 to 3.4 nm, with a more circular tendency. In fact, making an approach, it was possible to notice that the whole image was full of nanocrystals with different orientations. In the figures inserted with diffraction patterns,

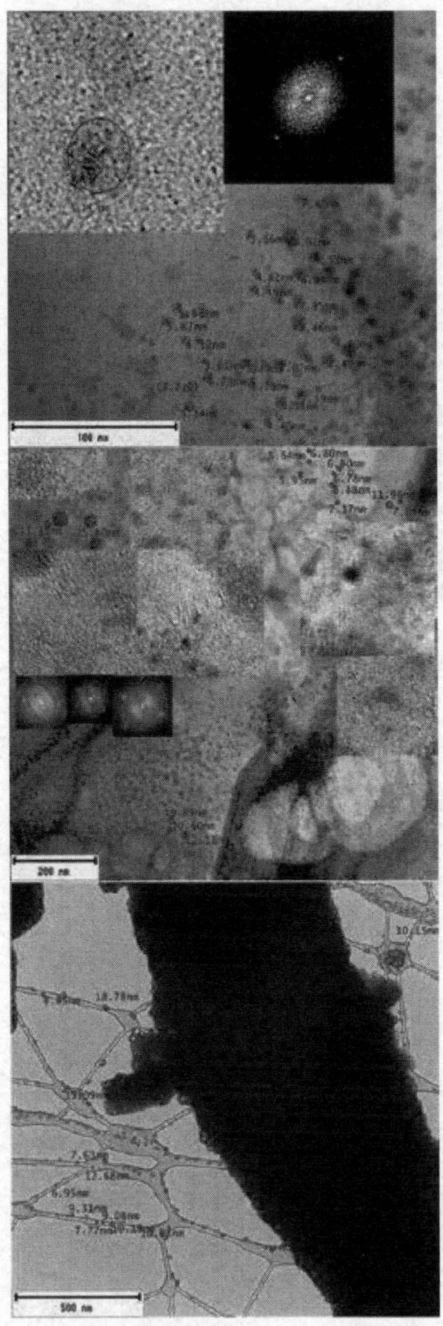

Figure 7. The HRTEM images of colloidal solution with acetone solvent.

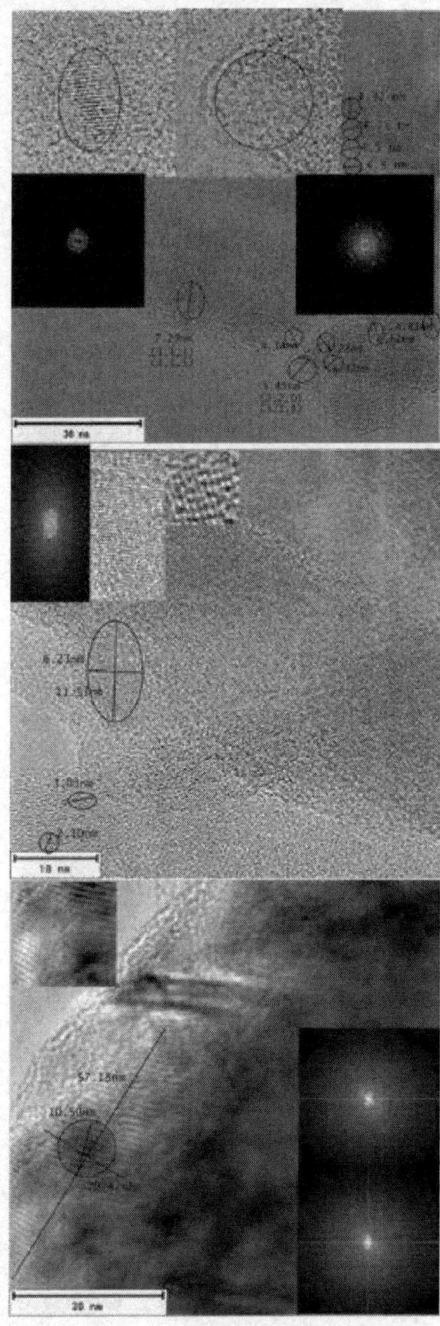

Figure 8. The HRTEM of the colloidal solution with ethanol solvent.

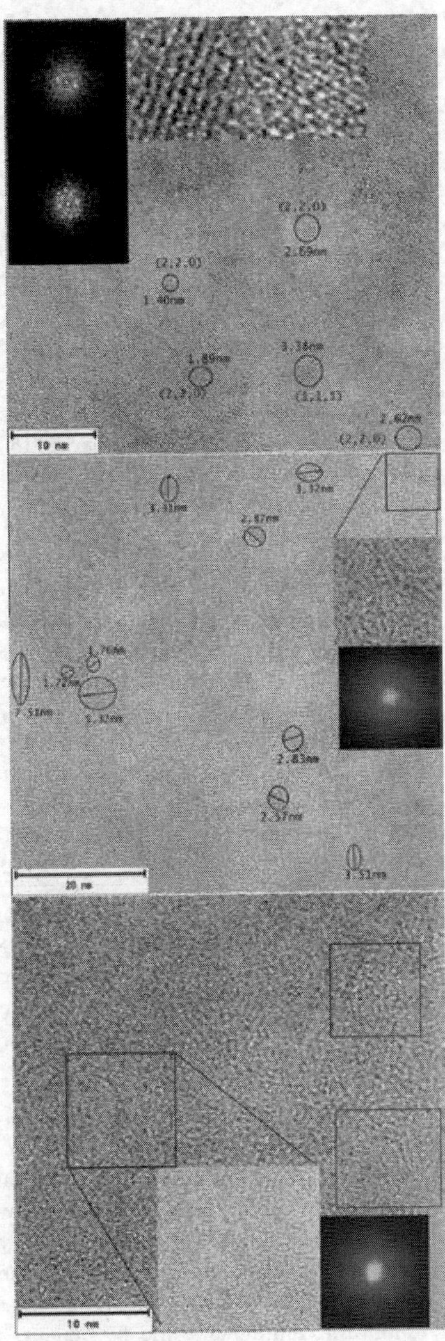

Figure 9. The HRTEM of the colloidal solution with methanol solvent.

they were found with crystalline preferential orientations of (220) and (111). With this image, we can verify and compare that the results of XRD are similar to those obtained with HRTEM, as well as with the results presented by Mallar et al. [57]. Even in the last image, one can see different regions with nanocrystals shape not well-defined, but with sizes of 1 nm or less.

The HRTEM images showed the approximate size of the nanocrystals (1.5 nm to 10 nm) and a roughly circular shape, Mallar et al. [57] according to their results mention that the HF apparently produces a decrease in the size of the nanocrystals; we can see this in the methanol sample because there are nanocrystals up to 1.4 nm. They also mention that there is a formation of spherical isolated crystals [64]. This is also compatible with our results, since these isolated crystals can be clearly seen in all the samples. According to the diffraction patterns shown in each of the HRTEM figures, some of these image insertions show almost continuous rings. This indicates that there are random orientations of neighboring crystals, that is, a polycrystalline composition. Such presence can also be noticed by diffuse halos which are indicators of an amorphous background that may be due to some type of silicon oxide or even amorphous silicon [57]. According to Mallar et al. [57], HF etching is more effective in the formation of isolated crystals, with regular shapes and nanometric sizes.

We can correlate the size of the nanocrystals with the solvent used. Such fact can be seen in the HRTEM images where the size of the nanocrystals decreased when the solvent was changed. More exactly, for the case of acetone solvent sample, the nanocrystal size was from 7 to 10 nm, with a large number of nanocrystals and core-shell structures. On the other hand, nanocrystal sizes between 4 and 6 nm were found for the ethanol sample, besides a reduction in the number of nanocrystals and amount of nanocrystals with sizes from 1 to 3 nm for the methanol sample was found. We attribute this event to the molar mass of each solvent; the molar mass of acetone is 58.08 g/mol, for ethanol we have a molar mass of 46.06 g/mol, and for methanol we have a molar mass of 32.02 g/mol. Due to the differences in the molar masses among the solvents, the formation of nanocrystals and agglomerates is possible.

Figure 10 shows histograms about the distribution of nanocrystals versus diameter. As can be seen, the sample contained in acetone possesses nanocrystals of larger size at 5 nm, while the sample contained in methanol has nanocrystals whose sizes are smaller than 5 nm. Finally, the ethanol sample contains intermediate nanocrystals whose size lies between the two mentioned samples. This can verify in a certain way, the proposed theory, the influence that the solvent has on the size of the nanocrystals.

In the article previously reported by the authors, photoluminescence spectra [54] were shown, which also correlate with PL intensity and nanocrystal size, where acetone samples (M_16 [54]) have a higher PL intensity, but a luminescent emission of longer wavelength; ethanol samples (M_14 [54]) had a lower PL intensity, but with a shift to lower wavelengths and methanol samples (M_09 [54]) had lower PL intensity, but with a higher energy luminescent emission than the previous samples. This can show that the solvent affects the size of the crystals, the luminescent emission and the emission energy. In this case, photoluminescence is affected by quantum confinement and by defects. The emission energy is originated by the nanocrystal size, while the emission intensity is attributed to the presence

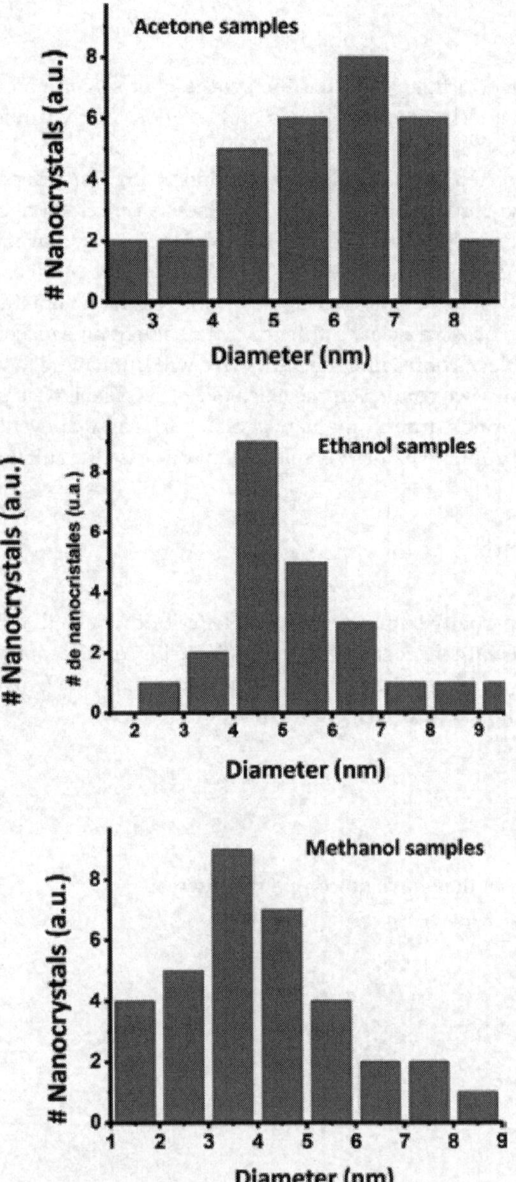

Figure 10. The histograms of the nanocrystals distribution vs. diameter.

of defects which are present in the nanocrystals due to the bonds that are formed with the solvent. From the HRTEM images, it was possible to verify that the different solvents affect the nanocrystal size.

4. Conclusions

We have demonstrated in this work that the synthesis of silicon nanocrystals by means of colloidal solutions is a technique very cheap and so good for controlling the Si-ncs size. By the XRD technique, it was possible to observe the Si-ncs obtained in the colloidal solutions. The AFM images depicted that the colloidal solutions with Si-ncs deposited on silicon have roughness, and it was observed that the Si-ncs possess a process of agglomerations, but it is indubitable that Si-ncs are present. The HRTEM images showed that colloidal solutions with Si-ncs have different agglomerations which exhibit a variable number of such nanocrystals depending on the solvent type. The Si-ncs sizes were possible to obtain in these images, and also, it was possible to observe a clear influence of the solvent type used in the solution. In this study, we were capable of controlling the Si-ncs size without having to use another technique more expensive. Finally, we confirmed the existence of core/shell type particles generated by the oxidated layers formed around the nanocrystal surface, such event may be controllable provided that the appropriate oxidant is selected, in this case the suitable solvent.

Acknowledgements

This work has been partially supported by CONACyT-255062 and VIEP-BUAP-LULJ EXC-2017, PFCE-2017. The authors acknowledge CUVVYT-BUAP laboratory for their help in the measurements of the samples. The authors also want to thank the University of Texas at San Antonio (UTSA) for the HRTEM measurements.

Conflict of interest

The authors declare that they have no competing interests.

References

[1] Richel A, Johnson NP, McComb DW. Observation of Bragg reflection in photonic crystals synthesized from air spheres in a titania matrix. Applied Physics Letters. 2000;**76**: 1816-1818. DOI: https://doi.org/10.1063/1.126175

[2] Zacharias M, Heitmann J, Scholz R, Kahler U, Schmidt M, Bläsing J. Size-controlled highly luminescent silicon nanocrystals: A SiO/SiO_2 superlattice approach. Applied Physics Letters. 2002;**80**:661

[3] Matsumoto T, Suzuki J, Ohnuma M, Kanemitsu Y, Masumoto Y. Evidence of quantum size effect in nanocrystalline silicon by optical absorption. Physical Review B. 2002; **63**:195322

[4] Gardelis S, Nassiopoulou AG, Manousiadis P, Milita S, Gkanatsiou A, Frangis N, Lioutas CB. Structural and optical characterization of two-dimensional arrays of Si nanocrystals embedded in SiO_2 for photovoltaic applications. Applied Physics Letters. 2012;**111**:083536

[5] Anderson IE, Rebecca A, Shircliff SB, Stradins P, Taylor PC, Collins RT. Synthesis and characterization of PECVD-grown, silane-terminated silicon quantum dots. In: 38th IEEE Photovoltaic Specialists Conference (PVSC). 2012. pp. 001890-001894

[6] Shaji N, Simmons CB, Thalakulam M, Klein LJ, Qin H, Luo H, Savage DE, Lagally MG, Rimberg AJ, Joynt R, Friesen M, Blick RH, Coppersmith SN, Eriksson MA. Spin blockade and lifetime-enhanced transport in a few-electron Si/SiGe double quantum dot. Nature Physics. 2008;**4**:540-544

[7] Nayfeh MH, Rao S, Nayfeh OM, Smith A, Therrien J. UV photodetectors with thin-film Si nanoparticles active medium. IEEE Transactions on Nanotechnology. 2005;**4**(Suppl 6): 660-668

[8] Zhang J, Liu J, Peng Q, Wang X, Li Y. Nearly monodisperse Cu_2O and CuO nanospheres: preparation and applications for sensitive gas sensors. Chemistry of Materials. 2006; **18**(Suppl 4):867-871

[9] Kim J, Grate JW. Single-enzyme nanoparticles armored by a nanometer-scale organic/ inorganic network. Nano Letters. 2003;**3**(Suppl 9):1219-1222

[10] Yang H, Wei W, Liu S. Monodispersed silica nanoparticles as carrier for co-immobilization of bi-enzyme and its application for glucose biosensing. Spectrochimica Acta. Part A, Molecular and Biomolecular Spectroscopy. 2014;**125**:183-188

[11] Chen HC, Qiu JT, Yang FL, Liu YC, Chen MC, Tsai RY, Yang HW, Lin CY, Lin CC, Wu TS, Tu YM, Xiao MC, Ho CH, Huang CC, Lai CS, Hua MY. Magnetic-composite-modified polycrystalline silicon nanowire field-effect transistor for vascular endothelial growth factor detection and cancer diagnosis. Analytical Chemistry. 2014;**86**(19):9443-9450

[12] Korzeniowska B, Woolley R, DeCourcey J, Wencel D, Loscher CE, McDonagh C. Intracellular pH-sensing using core/shell silica nanoparticles. Journal of Biomedical Nanotechnology. Jul 2014;**10**(7):1336-1345

[13] Deng H, Li X, Peng Q, Wang X, Chen J, Li Y. Monodisperse magnetic single-crystal ferrite microspheres. Angewandte Chemie, International Edition. 2005;**44**(Suppl 18):2782-2785

[14] Fronthal F, Trifonov T, Rodríguez A, Goyes C, Marsal LF, Borrull JF, Pallarès J. Electrical and optical characterization of porous silicon/p- crystalline silicon heterojunction diodes. AIP Conference Proceedings. 2008;**992**:780

[15] Izabela J, Filipek K, Duerinckx F, Kerschaver EV, Nieuwenhuysen KV, Beaucarne G, Poortmans J. Chirped porous silicon reflectors for thin-film epitaxial silicon solar cells. Journal of Applied Physics. 2008;**104**:073529

[16] Tsybeskov L, Duttagupta SP, Hirschman KD, Fauchet PM. Stable and efficient electroluminescence from a porous silicon-based bipolar device. Applied Physics Letters. 1996;**68**:2058

[17] Peng C, Hirschman KD, Fauchet PM. Carrier transport in porous silicon light-emitting devices. Journal of Applied Physics. 1996;**80**:295

[18] Perez EX, García FJ, Fenollosa R, Meseguer F. Photonic binding in silicon-colloid microcavities. Physical Review Letters. 2009;**103**:103902

[19] Tiwari S, Rana F, Chan K, Shi L, Hanafi H. Single charge and confinement effects in nano-crystal memories. Applied Physics Letters. 1996;**69**(9):1232

[20] Tiwari S, Rana F, Hanafi H, Hartstein A, Crabbe EF, Chan K. A silicon nanocrystals based memory. Applied Physics Letters. 1996;**68**(10):1377

[21] Nassiopoulou AG. In: Nalwa HS, editor. Encyclopedia of Nanoscience and Nanotechnology. Valencia, CA: American Scientific; 2004;**9**:793-813

[22] Nassiopoulou AG, Olzierski A, Tsoi E, Salonidou A, Kokonou M, Stoica T, Vescan L. Laterally ordered 2-D arrays of Si and Ge nanocrystals within SiO_2 thin layers for application in non-volatile memories. International Journal of Nanotechnology. 2009;**6**:18-34

[23] Green MA. Third Generation Photovoltaics: Advanced Solar Energy Conversion. Berlin, Heidelberg: Springer; 2003

[24] Conibeer G. Third-generation photovoltaics. Materials Today. 2007;**10**(11):42-50

[25] Rölver R, Berghoff B, Batzner D, Spangenberg B, Kurz H, Schmidt M, Stegemann B. Si/SiO_2 multiple quantum wells for all silicon tandem cells: Conductivity and photocurrent measurements Thin Solid Films 2006;**516**:6763

[26] Cho E-C, Park S, Hao X, Song D, Conibeer G, Park S-C, Green MA. Silicon quantum dot/crystalline silicon solar cells. Nanotechnology. 2008;**19**:245201

[27] Ficcadenti M, Pinto N, Morresi L, Murri R, Serenelli L, Tucci M, Falconieri M, Krasilnikova Sytchkova A, Grilli ML, Mittiga A, Izzi M, Pirozzi L, Jadkar SR. Si quantum dots for solar cell fabrication. Materials Science & Engineering, B: Advanced Functional Solid-State Materials. 2009;**66**:159-160

[28] Kim S-K, Cho C-H, Kim B-H, Park S-J, Lee JW. Electrical and optical characteristics of silicon nanocrystal solar cells. Applied Physics Letters. 2009;**95**:143120

[29] Stupca M, Alsalhi M, Saud TA, Almuhanna A, Nayfeh MH. Enhancement of polycrystalline silicon solar cells using ultrathin films of silicon nanoparticle. Applied Physics Letters. 2007;**91**:063107

[30] Yuan Z, Pucker G, Marconi A, Sgrignuoli F, Anopchenko A, Jestin Y, Ferrario L, Bellutti P, Pavesi L. Solar Energy Materials & Solar Cells. 2011;**95**(4):1224-1227

[31] Green MA, Conibeer G, Cho E-C, Konig D, Huang S, Song D, Scardera G, Cho Y-H, Fangsuwannarak T, Huang Y, Pink E, Bellet D, Bellet Amalric E, Puzzer T. In: 22nd European Photovoltaic Solar Energy Conference (Milan, Italy); 2007. pp. 1-4

[32] Igarashi M, Budiman MF, Pan W, Weiguo H, Tamura Y, Syazwan ME, Usami N, Samukawa S. Effects of formation of mini-bands in two-dimensional array of silicon nanodisks with SiC interlayer for quantum dot solar cells. Nanotechnology. 2013;**24**:015301

[33] Rodriguez I, Fenollosa R, Meseguer F. Silicon microspheres as UV, visible and infrared filters for cosmetics. Cosmetic & Toiletries. 2010;**125**(9):42-50

[34] Levitsky IA, Euler WB, Tokranova N, Rose A. Fluorescent polymer-porous silicon microcavity devices for explosive detection. Applied Physics Letters. 2007;**90**:041904

[35] Zhou Z, Brus L, Friesner R. Electronic structure and luminescence of 1.1- and 1.4-nm silicon nanocrystals: Oxide shell versus hydrogen passivation. Nano Letters. 2003;**3**:163-167

[36] Fojtik A, Henglein A. Luminescent colloidal silicon particles. Chemical Physics Letters. 1994;**221**:363-367

[37] Chen XY, Lu YF, Wu YH, Cho BJ, Liu MH, Dai DY, Song WD. Mechanisms of photoluminescence from silicon nanocrystals formed by pulsed-laser deposition in argon and oxygen ambient. Journal of Applied Physics. 2003;**93**:6311

[38] Vanhellemont J, De Gryse O, Clauws P. Critical precipitate size revisited and implications for oxygen precipitation in silicon. Journal of Applied Physics. 2005;**86**:221903

[39] Shimizu-Iwayama T, Fujita K, Nakao S, Saitoh K, Fujita T, Itoh N. Visible photoluminescence in Si+-implanted silica glass. Journal of Applied Physics. 1994;**75**:7779-7783

[40] Min KS, Shcheglov KV, Yang CM, Atwater HA, Brongersma ML, Polman A. Defect-related versus excitonic visible light emission from ion beam synthesized Si nanocrystals in SiO_2. Applied Physics Letters. 1996;**69**:2033-2035

[41] Littau KA, Szajowski PJ, Muller AJ, Kortan AR, Brus LE. A luminescent silicon nanocrystal colloid via a high-temperature aerosol reaction. The Journal of Physical Chemistry. 1993;**97**:1224-1230

[42] Holunga DM, Flagan RC, Atwater HA. A scalable turbulent mixing aerosol reactor for oxide-coated silicon nanoparticles. Industrial & Engineering Chemistry Research. 2005; **44**:6332-6341

[43] Zhang Q, Bayliss SC, Hutt DA. Blue photoluminescence and local structure of Si nanostructures embedded in SiO_2 matrices. Applied Physics Letters. 1995;**66**:1977-1979

[44] Schmidt JU, Schmidt B. Investigation of Si nanocluster formation in sputter-deposited silicon sub-oxides for nanocluster memory structures. Materials Science and Engineering B. 2003;**101**:28-33

[45] Lin CF, TsengW T, Feng MSJ. Formation and characteristics of silicon nanocrystals in plasma-enhanced chemical vapor-deposition silicon-rich oxide. Applied Physics. 2000; **87**:2808-2815

[46] Nozaki T, Sasaki K, Ogino T, Asahi D, Okazaki K. Microplasma synthesis of tunable photoluminescent silicon nanocrystals. Nanotechnology. 2007;**18**:235603

[47] Saunders WA, Sercel PC, Lee RB, Atwater H, Vahala KJ, Flanagan RC, Escorsi-Aparcio EJ. Synthesis of luminescent silicon clusters by spark ablation. Applied Physics Letters. 1993;**63**:1549-1551

[48] Kahler U, Hofmeister H. Visible light emission from Si nanocrystalline composites via reactive evaporation of SiO. Optical Materials. 2001;**17**:83-86

[49] Lin CF, TsengW T, Feng MS. Formation and characteristics of silicon nanocrystals in plasma-enhanced chemical-vapor-deposited silicon-rich oxide. Journal of Applied Physics. 2000;**87**:2808-2815

[50] Nozaki T, Sasaki K, Ogino T, Asahi D, Okazaki K. Microplasma synthesis of tunable photoluminescent silicon nanocrystals. Nanotechnology. 2007;**18**:235603

[51] Lioutas C, Vouroutzis N, Tsiaoussis I, Frangis N, Gardelis S, Nassiopoulou AG. Columnar growth of ultra-thin nanocrystalline Si films on quartz by low pressure chemical vapor deposition: accurate control of vertical size. Physica Status Solidi A: Applications and Materials Science. 2008;**205**:2615

[52] Shimizu-Iwayama T, Fujita K, Nakao S, Saitoh K, Fujita T, Itoh N. Visible photoluminescence in Si+-implanted silica glass. Journal of Applied Physics. 1994;**75**:7779-7783

[53] Min KS, Shcheglov KV, Yang CM, Atwater HA, BrongersmaM L, Polman A. Defect-related versus excitonic visible light emission from ion beam synthesized Si nanocrystals in SiO_2. Applied Physics Letters. 1996;**69**:2033-2035

[54] López JAL, Román AG, Barojas EG, Flores Gracia JF, Juárez JM, López JC. Synthesis of colloidal solutions with silicon nanocrystals from porous silicon. Nanoscale Research Letters. 2014;**9**:571

[55] Morris MC, McMurdie HF, Evans EH, Paretzkin B, de Groot JH, Hubbard CR, Carmet SJ, Standard X-ray Diffraction Powders Ptterns, 1976; section 13:44:25/sec 12:35

[56] Han PG, Poon MC, Sin KO, Wong M. Photoluminescent porous polycrystalline silicon. In Electron Devices Meeting, Proceedings IEEE Hong Kong. 1995:2-5

[57] Ray M, Hossain SM, Klie RF, Banerjee K, Ghosh S. Free standing luminescent silicon quantum dots: evidence of quantum confinement and defect related transitions. Nanotechnology. 2010;**21**(Suppl 50):9

[58] Zang JB, Wang YH, Zhao SZ, Bian LY, Lu J. Electrochemical properties of nanodiamond powder electrodes. Diamond and Related Materials. 2007;**16**:16-20

[59] He D, Shao L, Gong W, Xie E, Xu K, Chen G. Electron transport and electron field emission of nanodiamond synthesized by explosive detonation. Diamond and Related Materials. 2000;**9**:1600-1603

[60] Yerci S, Serincan U, Dogan I, Tokay S, Genisel M, Aydinli A, Turan R. Formation of silicon nanocrystals in sapphire by ion implantation and the origin of visible photoluminescence. Journal of Applied Physics. 2006;**100**:074301

[61] Boer EA, Brongersma ML, Atwater HA, Flagan RC, Bell LD. Localized charge injection in SiO_2 films containing silicon nanocrystals. Applied Physics Letters. 2001;**79**:791

[62] Maqbool M, Ali G, Cho SO, Ahmad I, Mehmood M, Kordesch ME. Nanocrystals formation and intense green emission in thermally annealed AlN:Ho films for microlaser cavities and photonic applications. Journal of Applied Physics. 2010;**108**:043528

[63] Cahay M, Garre K, Wu X, Poitras D, Lockwood DJ, Fairchild S. Physical properties of lanthanum monosulfide thin films grown on (100) silicon substrates. Journal of Applied Physics. 2006;**99**:123502

[64] Jasmin A, Rillera H, Semblante O, and Sarmago R. Surface morphology, microstructure, Raman characterization and magnetic ordering of oxidized Fe-sputtered films on silicon substrate. AIP Conference Proceedings. 2012;**1482**:572

Chapter 5

Twin Deformation Mechanisms in Nanocrystalline and Ultrafine-Grained Materials

Abstract

The review of the theoretical models, which describes mechanisms of deformation twinning in nanocrystalline and ultrafine-grained materials, is presented. Realization of special mechanisms of nanoscale deformation twin generation at locally distorted grain boundaries (GBs) in nanocrystalline and ultrafine-grained materials is observed. In particular, the micromechanisms of deformation twin formation occur through (1) the consequent emission of partial dislocations from GBs; (2) the cooperative emission of partial dislocations from GBs; and (3) the generation of multiplane nanoscale shear at GBs. The energy and stress characteristics of the deformation nanotwin generation at GBs in nanocrystalline and ultrafine-grained materials are calculated and analyzed. Competition between the twin generation mechanisms in nanocrystalline and ultrafine-grained materials is discussed.

Keywords: nanocrystalline and ultrafine-grained materials, nanotwins, plastic deformation, grain boundaries, dislocations

1. Introduction

At present, the study of the plastic behavior of nanostructured solids is one of the most important and rapidly developing directions in the mechanics of deformed solid and in the physics of condensed state. Nanostructured solids have unique physical, mechanical and chemical properties and are of great interest, both for fundamental and applied research [1]. For example, the strength and hardness of nanostructured materials are several times higher than those of conventional coarse-grained analogues of the same chemical composition. However, most nanocrystalline materials show low tensile ductility, which are highly undesirable for their practical applications. Increasing the ductility and fracture toughness of nanomaterials is a very important task, the solution of which can significantly expand the field of their application. With a large volume fraction occupied by GBs which act as effective obstacles for lattice dislocation slip (the dominant deformation mechanism in conventional coarse-grained polycrystals), the conventional lattice dislocation slip is hampered in nanostructured materials. At the same time, the specific features of the structure of nanocrystalline materials provide the action of specific deformation mechanisms, and the effect of which in coarse-grained materials was not observed or was insignificant. Identification of these specific mechanisms of plastic deformation is a key problem for understanding the nature of ductility and fracture toughness of nanostructured solids. According to modern concepts of plastic flow processes, the following specific mechanisms of plastic deformation act in nanocrystalline and ultrafine-grained materials: GB sliding [2, 3], rotational deformation mode [4–6], GBs migration [7–9] and deformation twinning [10–13]. The analysis of experimental investigations of deformation mechanisms allows us to formulate the main difference between nanocrystalline materials exhibiting low and high ductility. The point is that each nanocrystalline sample consists of a number of structural elements: grains of different sizes, GBs of various types and misorientations. In this case, several mechanisms of plastic deformation can act simultaneously in a nanocrystalline sample under mechanical loading. In general, different mechanisms of plastic deformation dominate in neighboring grains of different sizes and adjacent GBs. In nanocrystalline materials with low ductility, different mechanisms of deformation act independently of each other, which lead to a substantial inhomogeneity of plastic deformation and can cause the nucleation and evolution of nanocracks. At the same time, in nanocrystalline materials exhibiting high ductility, different mechanisms of plastic deformation effectively interact with each other. Intensive crossovers occur between different deformation mechanisms which accommodate the inhomogeneities of plastic deformation. One of the main specific deformation modes which contribute greatly to plastic flow in nanocrystalline and ultrafine-grained materials is considered deformation twinning mechanism. Following numerous experimental data, computer simulations and theoretical models [10–17], nanoscale twin deformation effectively operates in nanomaterials with various chemical compositions and structures. In doing so, in contrast to coarse-grained polycrystals where deformation twins are typically generated within grain interiors, in nanomaterials under mechanical load, twins are often generated at GBs; see [12] and references therein. In order to explain this experimentally documented fact indicative of specific deformation behavior of nanomaterials, it was suggested that nanoscale deformation twinning occurs through consequent emission of partial dislocations from GBs [10–13]. However, in this situation, partial dislocations should exist on every slip plane or be transformed from pre-existent GB dislocations which is hardly possible in real materials [13]. In order to avoid the discussed discrepancy, Zhu and coworkers [13] suggested new micromechanism of partial dislocation multiplication which realized due to successive processes of dislocation reactions and cross-slips providing existence of the partial dislocations at a GB on every slip plane. Thus, further consequent emission of such partial dislocations from GB can provide a nanoscale formation at GB [13]. At the same time, this approach operates with dislocation reactions each transforming a partial dislocation into two dislocations: a full dislocation and another partial dislocation. Such reactions are specified by very large energy barriers (being around the energy of a full dislocation), and thereby they are hardly typical in real materials. In order to respond to these questions, in theoretical works [14, 15], alternative mechanism of nanoscale twin formation at locally distorted GBs in deformed nanomaterials was suggested. According to results of the theoretical works [14, 15], GB dislocation can exist at locally distorted GB on every slip plane due to preceding plastic deformation and thereby cause nanoscale twin formation at GB. Taking this approach into account [14, 15], micromechanisms of deformation nanotwin formation can occur through (1) the consequent emission of partial dislocation from locally distorted GBs; (2) the cooperative emission of partial dislocations from locally distorted GBs; and (3) the generation of multiplane nanoscale shear at locally distorted GBs. Realization of these mechanisms is discussed in the next sections.

2. Mechanisms of deformation twin generation at locally distorted grain boundaries in nanocrystalline and ultrafine-grained materials

In this chapter, theoretical description of deformation twin generation mechanisms is based on results of the following theoretical papers [14–17]. According to these papers [14–17], generation of nanotwins occurs at locally distorted GB segments (GB segments being rich in GB dislocations) which were produced due to either events of consequent trapping of extrinsic lattice dislocations by GB and their splitting transformations into a wall of climbing GB dislocations (**Figure 1a–d**) or GB deformation processes involving slip and climb of GB dislocations (**Figure 1e–h**). The splitting of extrinsic dislocations at high-angle GBs is a well experimentally documented process [18] resulting at its initial stage in the formation of several closely located GB dislocations (**Figure 1a**). These processes allow GB dislocations to exist on almost every slip plane and thereby form a nanowall of GB dislocations (**Figure 1d**). In this situation, under action of external shear stress, a head dislocation of a pile-up is trapped by GB and splits into the GB dislocations (**Figure 1a** and **b**). After this process, the second lattice dislocation of the pile-up moves to and is trapped by GB where this dislocation splits into new GB dislocations (**Figure 1c**). In this case, after the splitting of the head dislocation of the pile-up, its second dislocation can reach the GB where this extrinsic dislocation splits into new GB dislocations (**Figure 1b** and **c**). Thus, consequent events of the splitting transformations of the head lattice dislocations forming pile-up into GB dislocations and climbing of these GB dislocations along GB can form a nanowall of GB dislocations located on almost every slip plane (**Figure 1d**). Both the transformation of GB dislocation into partial dislocations and emission of partial dislocation into grain interior are capable of producing a deformation nanotwin (for details, see below).

As follows from works [14, 15], formation of local distorted segments of GBs can be associated with GB plastic deformation processes. First, a nanostructured specimen is deformed by GB sliding that produces pile-ups of GB dislocations stopped by triple junctions of GBs (**Figure 1e**). Under the action of the external shear stress, the head GB dislocations of the pile-up split at triple junction and climb along GB (**Figure 1f–h**). As a result, a wall of climbing GB dislocations located on every (or almost every) slip plane is formed (**Figure 1h**). In general, local GB fragments being rich in GB dislocations can be formed at GBs "globally" distorted by plastic deformation. Such GBs are typical structural elements of bulk nanostructured materials

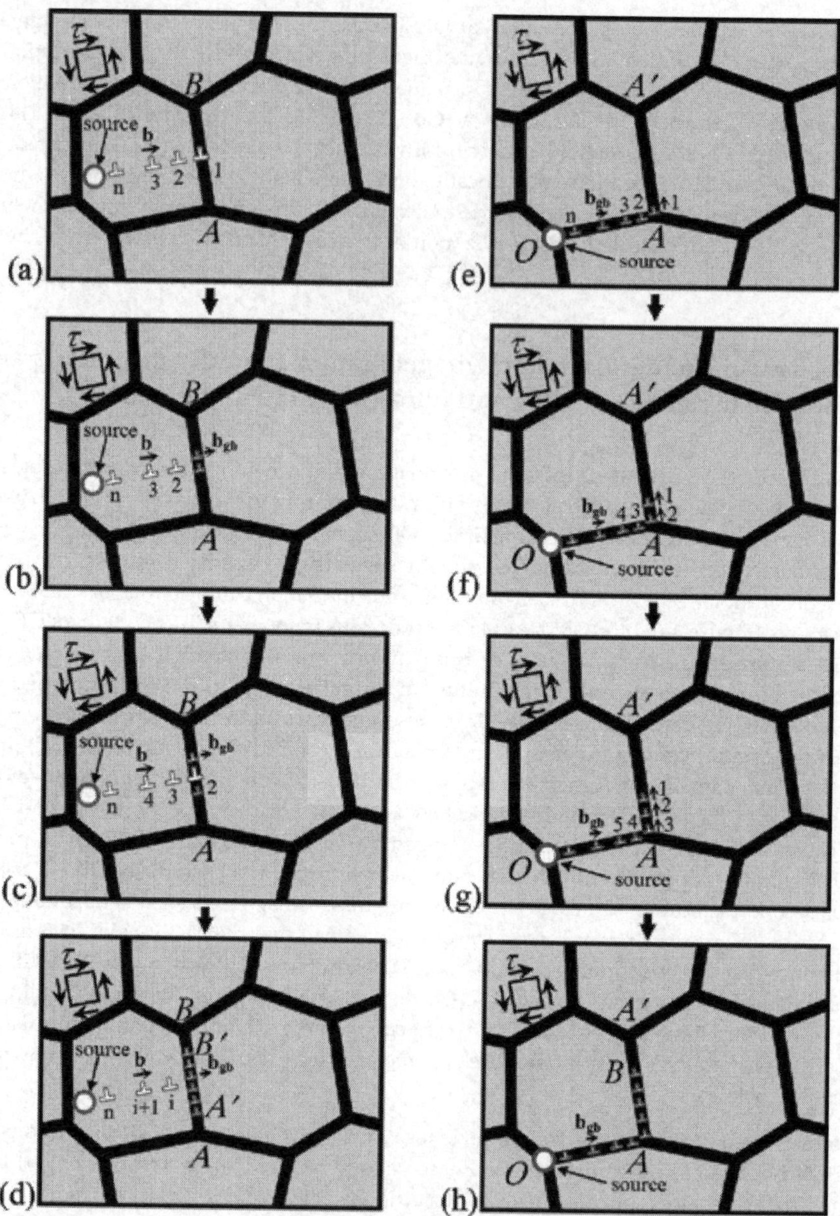

Figure 1. Mechanisms of formation of locally distorted GBs (grain boundaries) in nanocrystalline specimen under mechanical load (schematically). (a)–(d) Formation of a nano-sized wall of extra GB dislocation $A'B'$ through successive splitting of the head dislocations belonging to the pile-up of lattice edge dislocations stopped by GB AB and climb process of GB dislocations along the GB AB. (e)–(h) Formation of a nano-sized wall of extra GB dislocation AB through successive splitting of the head dislocations belonging to the pile-up of GB dislocations stopped by GB AA' and climb process of GB dislocations along the GB AA'.

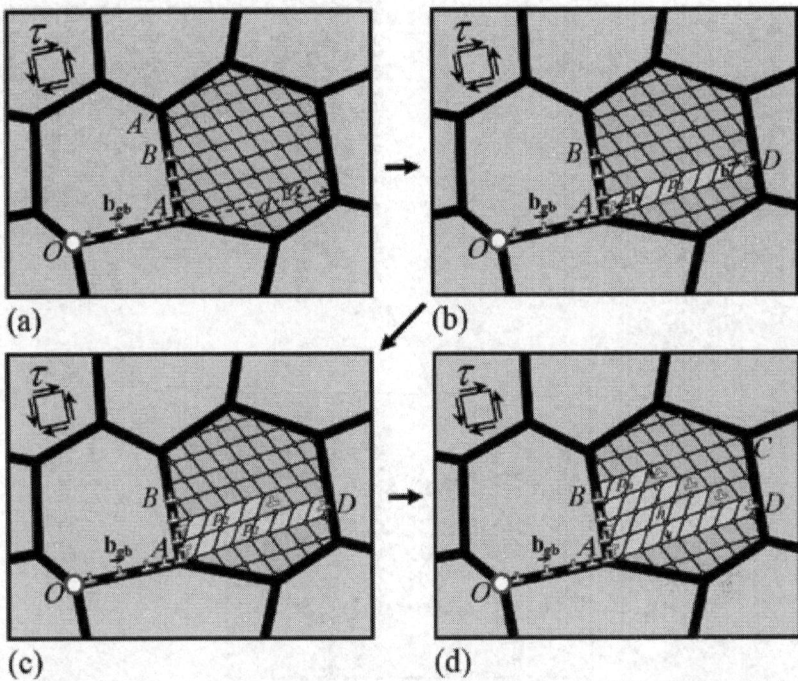

Figure 2. Mechanism of formation of nanoscale twins at locally distorted GBs (grain boundaries) through successive emission of partial dislocations in a nanocrystalline specimen under mechanical load (schematically). (a)–(d) Processes of successive dislocation emission from locally distorted GB fragment *AB* that move in adjacent grain interior and form a nanoscale twin *ABCD*.

fabricated by severe plastic deformation methods, and they can contain nanoscale fragments with GB dislocations located on every slip plane.

Thus, micromechanisms of nanotwin formation at locally distorted GB segments represent: (1) the consequent emission of partial dislocations from GBs; (2) the cooperative emission of partial dislocations from GBs; and (3) the generation of multiplane nanoscale shear at GBs. The former two micromechanisms of nanoscale twins generation occur through splitting of the GB dislocations into immobile GB dislocations and mobile partial dislocations (**Figures 2** and **3**). Consequent (**Figure 2**) or cooperative (**Figure 3**) gliding of the mobile dislocations along neighboring slip planes in a grain interior results in formation of a nanotwin.

Note that an energy barrier specifying the transformation of a GB dislocation at a local distorted GB segment into another GB dislocation and a partial dislocation (**Figures 2** and **3**) is around the energy of a partial dislocation. Thus, this barrier is lower than the barrier required for multiplication of partial dislocations (being around the energy of a full dislocation) considered by Zhu and coworkers [13]. In these circumstances, the splitting transformation (**Figures 2** and **3**) is more energetically favored as compared to the multiplication reaction.

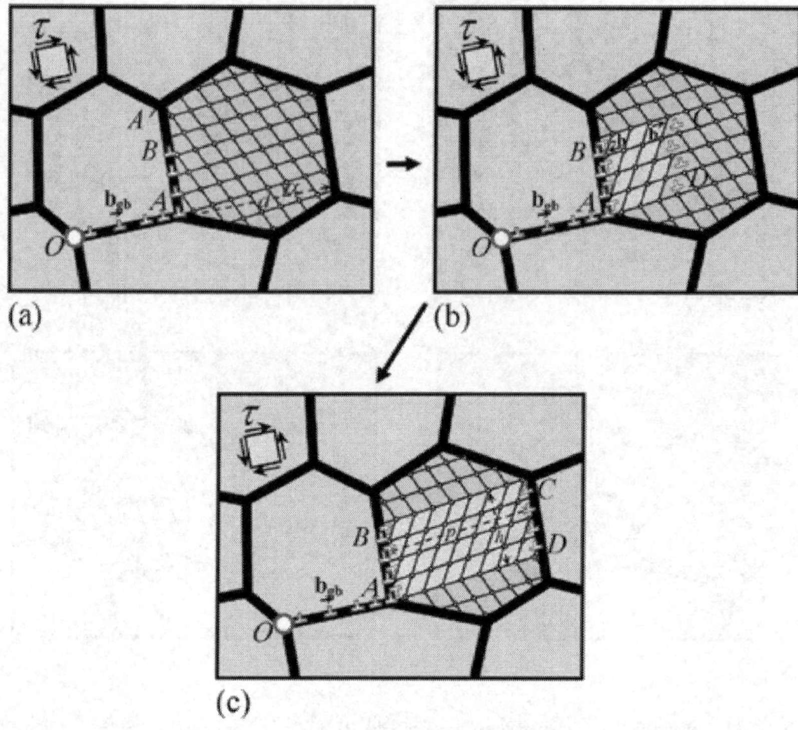

Figure 3. Mechanism of formation of nanoscale twins at locally distorted GBs (grain boundaries) through cooperative emission of partial dislocations in a nanocrystalline specimen under mechanical load (schematically). (a)–(c) Partial dislocation cooperatively emit from locally distorted GB fragment *AB* and move in adjacent grain interior toward the opposite GB forming a nanoscale twin *ABCD*.

The third mechanism for nanotwin formation at a locally distorted GB is multiplane nanoscale shear (**Figure 4**) firstly defined in Letter [19]. Following [19], a multiplane nanoscale shear is an ideal (rigid body) shear occurring simultaneously along several neighboring crystallographic planes within a nanoscale region—a three-dimensional region having two or three nanoscopic sizes—in a crystalline solid (this notion is based on that of multiplane ideal shear in infinite crystals [20]). The multiplane shear is characterized by the shear magnitude s (which is identical at any time moment, for all the planes where the shear occurs) gradually growing from 0 to the partial dislocation and geometric sizes of the nanoscale region where the shear occurs. For certain value of s, a nanotwin is generated within the region in question (**Figure 4**). Let us discuss in more detail geometric and energetic characteristics of the three mechanisms for nanotwin formation at locally distorted GBs.

2.1. Nanotwin formation due to consequent emission of partial dislocations

Figure 2 illustrates geometric features of nanotwin formation at locally nonequilibrium GBs in nanomaterials in the situation where local GB fragments with extra GB dislocations are formed

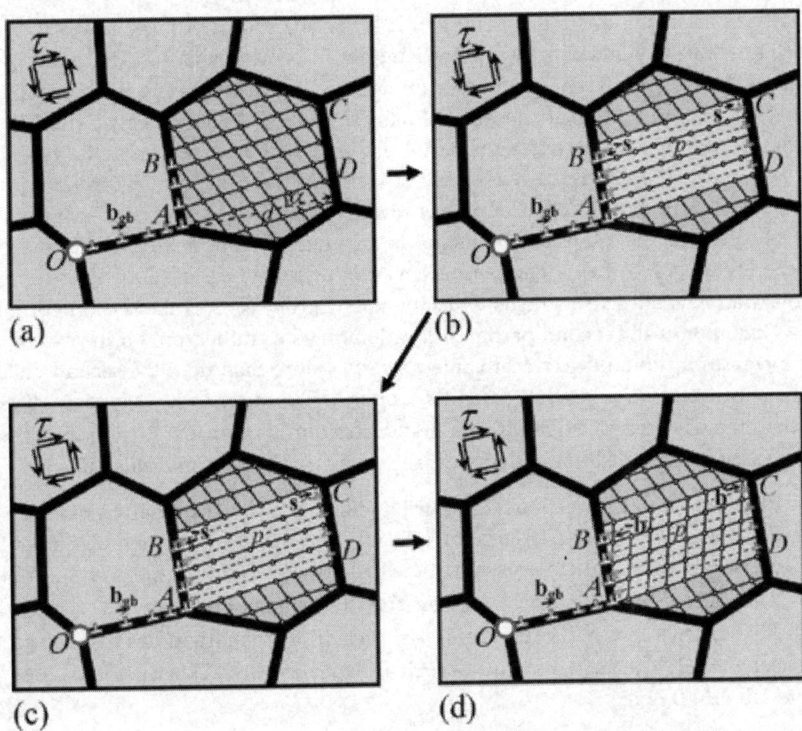

Figure 4. Mechanism of formation of nanoscale twins at locally distorted GBs (grain boundaries) through nanoscale multiplane shear in a nanocrystalline specimen under mechanical load (schematically). (a)–(d) Generation of nanotwin *ABCD* through subsequent transformation of non-crystallographic dislocations with Burgers vectors $\pm\mathbf{s}$ into twinning partial dislocations.

due to both GB sliding and stress-driven climb of GB dislocations. As a result, a nanoscale wall configuration *AB* of GB dislocations is formed at GB *AA′* (**Figure 2a**). In the framework of the model, the GB dislocations forming the wall of climbing dislocations transform into immobile GB dislocations (staying at the GB *AA′*) and mobile partial dislocation which are emitted from the GB *AA′* and can move along neighboring slip planes {111} in an adjacent grain (**Figure 2b–d**). In terms of the continuum approach, the emission of partial dislocations can be represented as formation of dislocation dipoles with Burgers vectors $\pm b$ and stacking faults (**Figure 2b–d**). Consequent generation of such dipoles of partial dislocations joined by stacking faults is capable of forming a nanoscale twin (**Figure 2b–d**). Such consequent events of partial dislocation emission from GBs were examined in several theoretical works (see, e.g., [16, 17]), which, however, did not concern formation of locally nonequilibrium GB structures considered here as initial ones for nanotwin generation.

The angle α specifies orientation of {111} slip planes for partial dislocations relative to the GB *AA′* plane (**Figure 2a**). The magnitude of Burgers vectors of partial dislocations is equal to $b = a/\sqrt{6}$. The distance δ is between the neighboring slip planes {111} and is related to the crystal lattice parameter a as follows $\delta = a/\sqrt{3}$. When the ith dislocation moves in the grain interior (**Figure 2b–d**), a stacking fault of the length p_i is formed behind it. The stacking fault is characterized by the specific energy (per its unit area) γ, which serves as a hampering force for the partial dislocation slip. The dislocation slip is driven by the shear stress τ. The first partial dislocation is emitted from the triple junction A (**Figure 2b**) and moves across the grain interior toward the opposite GB (**Figure 2b**) when the shear stress reaches its critical value of τ_{c1}. After emission, the first partial dislocation moves toward opposite GB and, depending on shear stress level τ, reaches the opposite GB or stops in the grain interior moving over some distance p_1 and creates the stress fields hampering emission of a new dislocation. The first emitted partial dislocation creates stress fields which hamper the emission of the second partial dislocation. As a corollary, the second partial dislocation may be emitted only if the external shear stress τ increases up to a new critical value $\tau_{c2} > \tau_{c1}$. More than that, the second dislocation under the shear stress τ_{c2} does not reach the opposite GB, but moves over some distance p_2 shorter than the distance p_1 moved by the first dislocation (**Figure 2c**). It is because the stress field created by the first dislocation hampers slip of the second partial dislocation.

Also, the discussed trends come into play during emission of other partial dislocations due to the effects of previously emitted dislocations. That is, the critical stress for emission of the nth dislocation is larger than that for emission of the $(n-1)$th dislocation ($\tau_{c(n)} > \tau_{c(n-1)}$), and this stress drives slip of the nth dislocation over the distance shorter than that moved by the $(n-1)$th dislocation ($p_n < p_{n-1}$) (**Figure 2**). As a result, the nanotwin has a shape schematically presented in **Figure 2d**. Nanotwins of such a shape have been experimentally observed in nanomaterials [13] (**Figure 5**).

To analyze the suggested model, we consider the energy characteristics of the nanoscale twin generation due to consequent emission of partial dislocations from locally nonequilibrium GBs (**Figure 2**). First, define the conditions which are necessary for the energetically favorable emission of the first partial dislocation, which can be represented as formation of a dipole AD of partial dislocations with Burgers vectors $\pm b$ (**Figure 2b**). Generation of this partial dislocation dipole is characterized by the energy change ΔW_1 (per unit length of the pile-up head dislocation) defined as $\Delta W_1 = \Delta W_{1,final} - W_{1,initial}$, where $W_{1,final}$ and $W_{1,initial}$ are the energies of the considered defect configuration in its final (**Figure 2b**) and initial (**Figure 2a**)

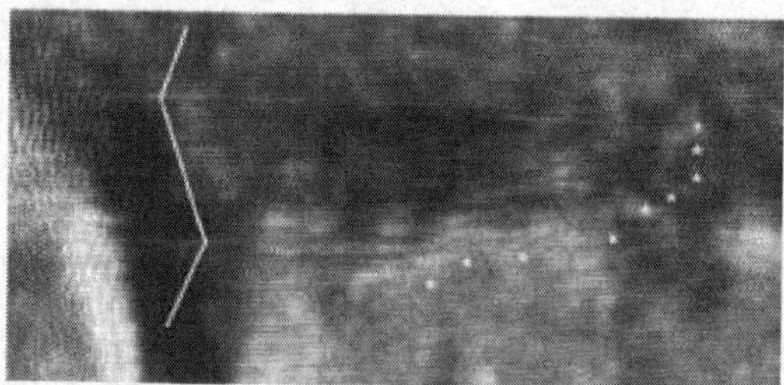

Figure 5. High resolution electron transmission microscopy image showing a deformation twin which ends in the grain interior as marked by the white asterisks in electrodeposited nanocrystalline Ni. Reprinted with permission from Ref. [13]. Copyright (2009), AIP Publishing LLC.

states, respectively. Formation of the dislocation dipole (**Figure 2b**) occurs as an energetically favorable process. if $\Delta W_1 < 0$. The energy change in question has the five terms:

$$\Delta W_1 = E^b + E^{\Delta - b} + E^{b - b_{gb}} + E_{\tau 1} + E_{\gamma 1}, \tag{1}$$

where E^b is the proper energy of the dipole of the Shockley partial dislocations having Burgers vectors $\pm b$; $E^{\Delta - b}$ is the energy that specifies the interaction between the partial dislocation dipole and the wall *AB* of GB dislocations; $E^{b - b_{gb}}$ is the energy that specifies the interaction between the partial dislocation dipole and the pile-up *OA* of GB dislocations; $E_{\tau 1}$ is the work spent by the external shear stress τ on movement of a mobile partial dislocation over the distance p; and $E_{\gamma 1}$ is the energy of the stacking fault formed between the partial dislocations belonging to the dipole *AD*.

Detailed description of all the terms figuring on the right-hand side of Eq. (1) is given in the theoretical paper [14, 15]. Using Eq. (1), we calculated the dependences of the energy change ΔW_1 on the distance p_1. We performed calculations for nanocrystalline nickel (Ni) and copper (Cu) for the following model parameter values characterizing Ni: G = 73 GPa, $\nu = 0.34$, $a = 0.352$ nm, $b_{gb} \approx 0.1$ nm, and $\gamma_{Ni} = 0.110$ J/m^2 [21] and Cu: G = 44 GPa, $\nu = 0.3$, $a = 0.358$ nm, $b_{gb} \approx 0.1$ nm, and $\gamma_{Cu} = 0.045$ J/m^2 [21]. The dependences $\Delta W_1(p_1)$ were calculated, for $d = 30$ nm, $n_c = 5$, $\tau = 100$ MPa and various values of α. The dependences $\Delta W_1(p_1)$ show the trend that emission of the first dislocation is enhanced when α decreases, and it is the most favorable at $\alpha = 0^{\circ}$.

Now let us consider the energy characteristics of emission of the nth partial dislocation, for $n > 1$. Emission of the nth partial dislocation is equivalent to formation of the nth dislocation dipole (**Figure 2**) in the nanocrystalline solid initially containing the dislocation-pile up and $(n-1)$ dipoles of partial dislocations is characterized by the energy change ΔW_n (per unit length of a partial dislocation) defined as $\Delta W_n = W_n - W_{n-1}$, where W_n and W_{n-1} are the energies of the considered defect configuration with n and $n-1$ partial dislocation dipoles, respectively. Formation of the nth dislocation dipole (**Figure 2**) occurs as an energetically favorable process, if $\Delta W_n < 0$. The energy change ΔW_n can be represented as follows:

$$\begin{aligned}\Delta W_n = {} & E_{\Sigma}^{b(n)} - E_{\Sigma}^{b(n-1)} + E_{\Sigma}^{c-b(n)} - E_{\Sigma}^{c-b(n-1)} + E_{\Sigma}^{\Delta - b(n)} - E_{\Sigma}^{\Delta - b(n-1)} \\ & + E_{\Sigma}^{b-b(n)} - E_{\Sigma}^{b-b(n-1)} + E_{\gamma\Sigma}^{(n)} - E_{\gamma\Sigma}^{(n-1)} + E_{\tau\Sigma}^{(n)} - E_{\tau\Sigma}^{(n-1)},\end{aligned} \tag{2}$$

where $E_{\Sigma}^{b(n-1)}$ and $E_{\Sigma}^{b(n)}$ are the total self-energies of $(n-1)$ and n partial dislocation dipoles, respectively; $E_{\Sigma}^{\Delta - b(n-1)}$ is the elastic interaction energy of $(n-1)$ partial dislocation dipoles with the wall *AB* of GB dislocations; $E_{\Sigma}^{\Delta - b(n)}$ is the elastic interaction energy of n partial dislocation dipoles with the wall *AB* of GB dislocations; $E_{\Sigma}^{c-b(n-1)}$ and $E_{\Sigma}^{c-b(n)}$ are the elastic interaction energies of the GB dislocation pile-up *OA* with $(n-1)$ and n partial dislocation dipoles, respectively; $E_{\Sigma}^{b-b(n-1)}$ and $E_{\Sigma}^{b-b(n)}$ are the sums of the energies specifying the dipole-dipole interaction in situations with $n-1$ and n dislocation dipoles, respectively; $E_{\gamma\Sigma}^{(n-1)}$ and

$E^{(n)}_{\gamma\Sigma}$ are the sums of the energies specifying the stacking faults in situations with $(n-1)$ and n dislocation dipoles, respectively; $E^{(n-1)}_{\tau\Sigma}$ and $E^{(n)}_{\tau\Sigma}$ are the sums of the energies specifying the interaction between the external shear stress τ as well as $(n-1)$ and n partial dislocation dipoles, respectively. Detailed description of all the terms figuring on the right-hand side of Eq. (2) is given in the theoretical papers [14, 15]. With the help of Eq. (2) for the energy change ΔW_n, we calculated dependences of ΔW_n on the distance p_n. With these dependences, we also calculated the critical shear stress $\tau_{c(n)}$ (that can be defined as the minimum stress at which emission of the nth partial dislocation from the GB occurs). The critical stress $\tau_{c(n)}$ is calculated from the equation $\Delta W_n(p_n = 1\text{nm}) = 0$. The dependences of critical shear stress $\tau_{c(n)}$ on nanotwin thickness are presented in **Figure 6**. As it follows from the dependences $\tau_{c(n)}(h)$, the critical shear stress $\tau_{c(n)}$ decreases when the grain size d increases and/or the nanotwin thickness h decreases. For instance, for d = 50 nm, the generation of a nanotwin having the thickness h = 3 nm occurs in Cu at the critical shear stress $\tau_{c(n)} \approx 2.2$ GPa (**Figure 6**). This value is very high, but it can be reached in shock load tests of nanocrystalline materials.

2.2. Nanotwin formation due to cooperative emission of partial dislocations

The second micromechanism of nanotwin formation is realized through cooperative emission of partial dislocations from locally distorted GBs in deformed nanomaterials. As in the previous case, the initial defect configuration represents a nanoscale wall of GB dislocations *AB* located on every (or almost every) slip plane (**Figure 3a**). In this situation, the GB dislocations cooperatively emit from GB and move together along neighboring slip plane forming a nanotwin (**Figure 3b** and **c**). This mechanism in the situation where the GB dislocations are located on every slip plane has been considered in theoretical papers [14, 15]. As a result, the nanotwin *ABCD* crosses the grain and joins two opposite GBs (**Figure 3c**). Such nanotwins have been experimentally observed in nanocrystalline nickel (Ni) [11] (**Figure 7**).

However, cooperative emission of partial dislocations from GBs also can occur in situation where the GB dislocations in the initial wall configuration *AB* are located on not all of slip planes. In this case, there are some gaps in arrangement of the GB b_{gb}-dislocations at GB fragment *AB*. Transformations of the GB b_{gb}-dislocations can occur in only a part of the set of

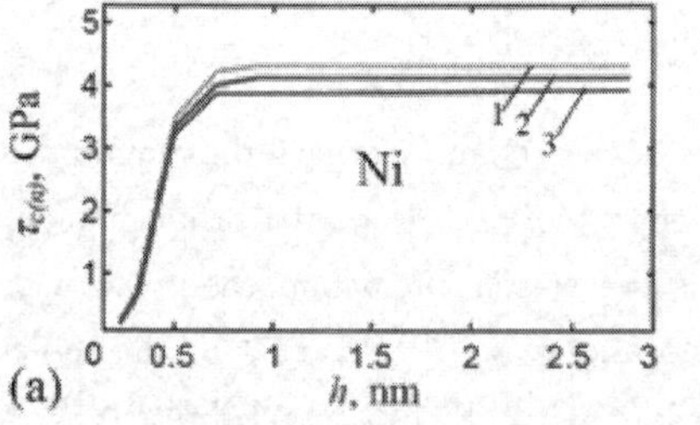

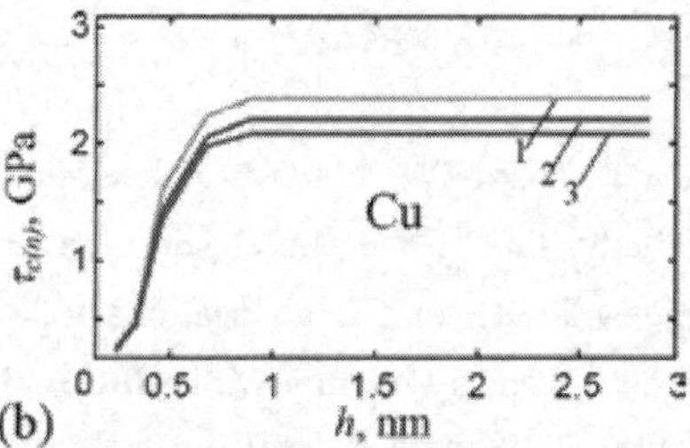

Figure 6. Dependences of the critical shear stress $\tau_{c(n)}$ on the nanotwin thickness h in the exemplary cases of nanocrystalline (a) nickel (Ni) and (b) copper (Cu), for various values of grain size d = 25 (curve 1), 50 (curve 2) and 100 (curve 3) nm.

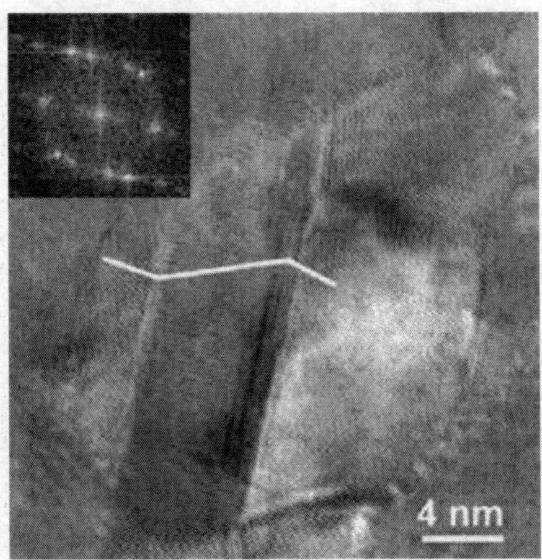

Figure 7. High resolution electron transmission microscopy image showing a deformation twin which crosses the grain interior and joins two opposite grain boundaries in nanocrystalline Ni. Reprinted from Ref. [11]. Copyright (2008), with permission from Elsevier.

crystallographic {111} planes adjacent to the GB fragment *AB* (**Figure 3**). Nevertheless, a nanotwin can be generated through cooperative emission of partial dislocations from such a GB fragment, if partial dislocations are generated on slip planes where the initial GB dislocations are absent.

Analyze the energy characteristics of the twin formation due to cooperative emission of dislocations from GB in nanomaterials (**Figure 3**). The cooperative emission process (**Figure 3**) is characterized by the energy difference $\Delta W'_n = W'_n - W'$, where W'_n and W' are the energies of the defect configuration in its final (**Figure 3b** and **c**) and initial (**Figure 3a**) states, respectively (after and before the emission process, respectively). In this situation, the nanotwin generation is energetically favorable, if $\Delta W'_n < 0$. The energy difference $\Delta W'_n$ has the six basic terms:

$$\Delta W'_n = E^{b}_{\Sigma n} + E^{c-b}_{\Sigma n} + E^{\Delta-b}_{\Sigma n} + E^{b-b}_{\Sigma n} + E^{\gamma}_{\Sigma n} + E^{\tau}_{\Sigma n}, \quad (3)$$

where $E^{b}_{\Sigma n}$ is the total self-energies of n partial dislocation dipoles; $E^{\Delta-b}_{\Sigma n}$ is the elastic interaction energy of n partial dislocation dipoles with the nanoscale wall *AB* of GB dislocations; $E^{c-b}_{\Sigma n}$ is the elastic interaction energy of the GB dislocation pile-up *OA* with n partial dislocation dipoles; $E^{b-b}_{\Sigma n}$ is the elastic energy of all the dipole-dipole interactions for n dislocation dipoles; E^{γ}_{n} is the energy of the twin boundaries; and $E^{\tau}_{\Sigma n}$ is the elastic interaction energy of the external shear stress τ with n partial dislocation dipoles.

Calculation of all the terms figuring on the right-hand side of Eq. (3) is given in the theoretical paper [14, 15]. With the help of Eq. (3) for the energy change $\Delta W'_n$, we revealed dependences of

$\Delta W'_n$ on the distance p moved by the n partial dislocation in bulk of grain. With these results, we also calculated the critical shear stress $\tau'_{c(n)}$ that is the minimum stress at which the cooperative emission of n partial dislocation from GB is energetically favorable (**Figure 3**). More precisely, the critical shear stress $\tau'_{c(n)}$ can be found from the conditions that $\Delta W'_n(p = p') = 0$ (where $p' = 1\text{nm}$), $\Delta W'_n|_{p>p'} < 0$, and $\left.\frac{\partial \Delta W'_n}{\partial p}\right|_{p>p'} \leq 0$.

Note that, if the inequalities $\Delta W'_n|_{p>p'} < 0$ and $\left.\frac{\partial \Delta W'_n}{\partial p}\right|_{p>p'} < 0$ are valid, the dependences of the energy change $\Delta W'_n(p)$ on the distance p moved by the partial dislocation group within a grain are monotonously decreasing and negatively valued functions. In this case, the group AB of n partial b-dislocations is generated from locally distorted GB AA' and move across the grain over the distance $p = d$ toward the opposite GB where it is stopped. As a corollary, a nanotwin is formed which joins the two opposite GBs (**Figure 3c**).

In another case, the inequalities $\Delta W'_n|_{p>p'} < 0$, $\left.\frac{\partial \Delta W'_n}{\partial p}\right|_{p=p_{eq}} = 0$ and $\left.\frac{\partial^2 \Delta W'_n}{\partial p^2}\right|_{p=p_{eq}} > 0$ are valid. In this case, a function $\Delta W'_n(p)$ has its minimum corresponding to the equilibrium distance p_{eq} moved by the group of the emitted partial b-dislocations or, in other words, the equilibrium position p_{eq} of the nanotwin front AB of the nanotwin $ABCD$ (**Figure 3b**).

Dependences of the critical shear stress $\tau'_{c(n)}$ on the nanotwin thickness h, for various values of n_c, are presented in **Figure 8**. As it is seen in **Figure 8**, values of the critical shear stresses $\tau'_{c(n)}$ decrease with raising the number n_c of GB dislocations in the pile-up OA (see **Figure 3**).

2.3. Nanotwin formation due to generation of nanoscale multiplane shear

The third mechanism of nanotwin formation is realized through the generation of nanoscale multiplane shear at locally distorted GBs in deformed nanomaterials (**Figure 4**). As it has been noted previously, nanoscale multiplane shear is defined in work [19] as a multiplane ideal shear occurring within a nanoscale region, a three-dimensional region having two or three nanoscopic sizes. For instance, nanoscale multiplane shear can occur and produce a twin

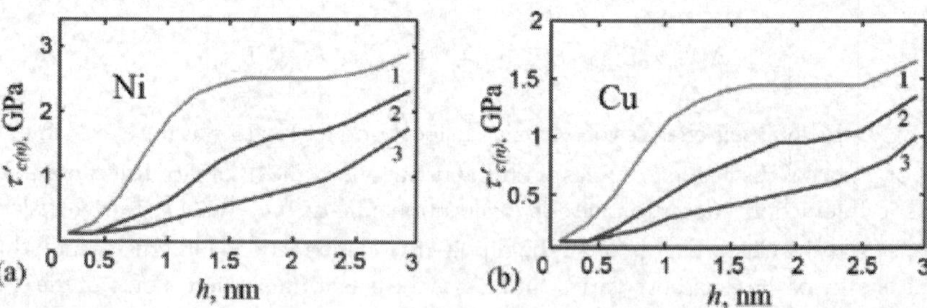

Figure 8. Dependences of the critical shear stress $\tau'_{c(n)}$ on the nanotwin thickness h in the exemplary cases of nanocrystalline (a) nickel (Ni) and (b) copper (Cu), for various values of grain boundary dislocations in the pile-up n_c = 4 (curve 1), 6 (curve 2) and 8 (curve 3).

within a nanoscale internal region of a grain of a deformed nanomaterial (**Figure 4**). More precisely, in the model situation, a deformation twin is produced under the action of a shear stress τ (**Figure 4**) through nanoscale multiplane shear or, in terms of the dislocation theory, through simultaneous nucleation of n dipoles of noncrystallographic dislocations with tiny Burgers vectors $\pm s$ (**Figure 4a–d**). The noncrystallographic dislocations of the dipoles are formed at opposite GB fragments, *AB* and *CD*, on adjacent {111} planes. The Burgers vectors $\pm s$ of all the dislocations are the same in magnitude and grow simultaneously from zero to the Burgers vectors $\pm b$ of partial dislocations during the nanotwin formation process. In doing so, since there are preexistent GB dislocations at the GB fragment *AB*, the noncrystallographic dislocations at the GB fragment *AB* merge with these preexistent GB dislocations. As a corollary, during the nanotwin formation process, evolution of the noncrystallographic dislocations at the GB fragment *AB* manifests itself in evolution of the GB dislocations at this fragment (**Figure 4b–d**).

In the case of fcc metals (in particular, copper (Cu)), the generated dipoles of noncrystallographic partial dislocations are formed in adjacent slip planes {111} assumed to be normal to the grain boundary fragments *AB* and *CD*. The region *ABCD* (a rectangle with sizes h and p) is subjected to nanoscale multiplane shear, which leads to the formation of a nanoscale twin within this region (**Figure 4**). In doing so, *AB* length = *CD* length = h; and *AD* length = *BC* length = p, where $p = d/\cos\alpha$ and d is the grain size (**Figure 4a**). As with the previously considered mechanisms for nanotwin formation, the distance δ between neighboring dipoles of Shockley dislocations is equal to the distance between the neighboring slip planes {111} and is in the following relationship with the crystal lattice parameter a: $\delta = a/\sqrt{3}$.

Analyze energetic characteristics of the generation of nanoscale twins through nanoscale multiplane shear initiated at locally distorted GBs in nanocrystalline materials (**Figure 4**). In this case, in terms of the dislocation theory, a deformation twin is produced under the action of a shear stress τ through simultaneous nucleation of n dipoles of noncrystallographic dislocations with tiny Burgers vectors $\pm s$ whose magnitudes $n = 3$ gradually grow from 0 to b during the nanotwin formation process (**Figure 4**).

In general, the energy change ΔW_N that characterizes the nanotwin generation through multiplane nanoscale shear (**Figure 4**) has the seven key terms:

$$\Delta W_N = E^{s}_{\Sigma n} + E^{c-s}_{\Sigma n} + E^{\Delta-s}_{\Sigma n} + E^{s-s}_{\Sigma n} + W_{interior}(s) + W_{AC-BD}(s) + A, \tag{4}$$

Here $E^{s}_{\Sigma n}$ is the total self-energies of n dipoles of noncrystallographic $\pm s$-dislocations; $E^{c-s}_{\Sigma n}$ is the elastic interaction energy of the pile-up of GB dislocations with n dipoles of noncrystallographic $\pm s$-dislocations; $E^{\Delta-s}_{\Sigma n}$ is the elastic interaction energy of the wall *AB* of GB dislocations with n dipoles of noncrystallographic $\pm s$-dislocations; $E^{s-s}_{\Sigma n}$ is the elastic energy of all dipole-dipole interactions for n dipoles of noncrystallographic $\pm s$-dislocations; $W_{interior}$ denotes the energy of the interior area of the plastically sheared nanocrystal *ABCD*; W_{AD-BC} is the energy of the interfaces, *AD* and *BC*, between the sheared region *ABCD* and the neighboring material; and A is the work of the external shear stress τ, spent to the plastic shear within the region *ABCD*.

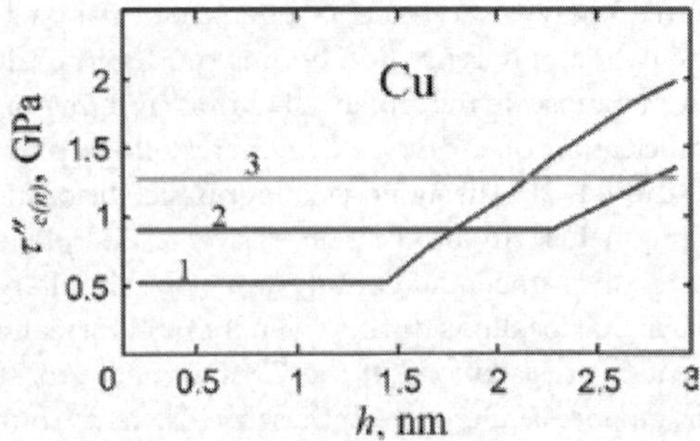

Figure 9. Dependences of the critical shear stress $\tau''_{c(n)}$ on the nanotwin thickness h in the exemplary cases of nanocrystalline copper (Cu), for various values of the grain size d = 25 (curve 1), 50 (curve 2) and 100 (curve 3) nm.

Calculations of all the terms figuring on the right-hand side of Eq. (4) are given in the theoretical paper [14, 15]. Based on these calculations in the exemplary cases of nanocrystalline Cu, we revealed dependences of the energy change ΔW_N on the Burgers vector magnitude s of the noncrystallographic dislocations for various sizes of the nanotwin. With functions $\Delta W_N(s)$, the critical shear stress $\tau''_{c(n)}$ (the minimum stress at which the nanotwin formation occurs) was calculated. As it follows from the dependences $\Delta W_N(s)$ the critical stress $\tau''_{c(3)} \approx 0.5$ GPa at $n = 3$, and $\tau''_{c(15)} \approx 2$ GPa at $n = 15$. The dependences of the critical shear stress $\tau''_{c(n)}$ on the nanotwin thickness h are presented in **Figure 9**, for various values of the grain size d. The dependences $\tau''_{c(n)}(h)$ show that the critical shear stress $\tau''_{c(n)}$ decreases when the grain size d increases and/or the nanotwin thickness h decreases (**Figure 9**).

3. Comparison of critical shear stress for nanotwin generation at grain boundaries due to various deformation mechanisms

In this section, we compare critical shear stresses for the considered mechanisms of nanotwin generation at locally distorted GBs in deformed nanomaterials. The dependences of the critical shear stresses $\tau_{c(n)}$ and $\tau'_{c(n)}$, for nickel (Ni), and $\tau_{c(n)}$, $\tau'_{c(n)}$ and $\tau''_{c(n)}$, for copper (Cu) on the nanotwin thickness h are presented in **Figure 10**. As it follows from **Figure 10**, for nickel (Ni), the cooperative emission of partial dislocations from GBs (**Figure 3**) is characterized by the lowest critical shear stress. In the case of Cu, the competition between different deformation mechanisms is more complicated. Therefore, for Cu, the mechanism of cooperative dislocation emission is realized at the lowest stress level $\tau'_{c(n)} < \tau''_{c(n)} < \tau_{c(n)}$ in the case of ultrathin nanotwin generation (with thickness $h < 1$ nm). At the same time, for Cu, the mechanism of nanoscale multiplane shear (**Figure 4**) occurs at the lowest stress level (**Figure 10**) in the range of nanotwin thickness h from 1 to 2 nm. For large nanotwins with thickness $h > 2$ nm, the mechanism of the cooperative emission of partial dislocations from GBs (**Figure 3**) is the most favorable process in copper again (**Figure 10**).

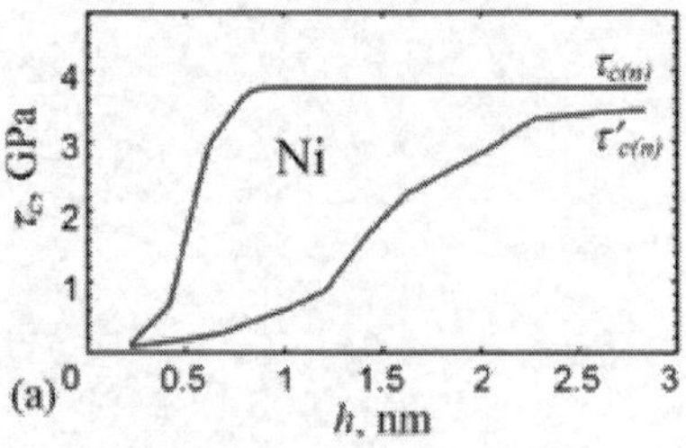

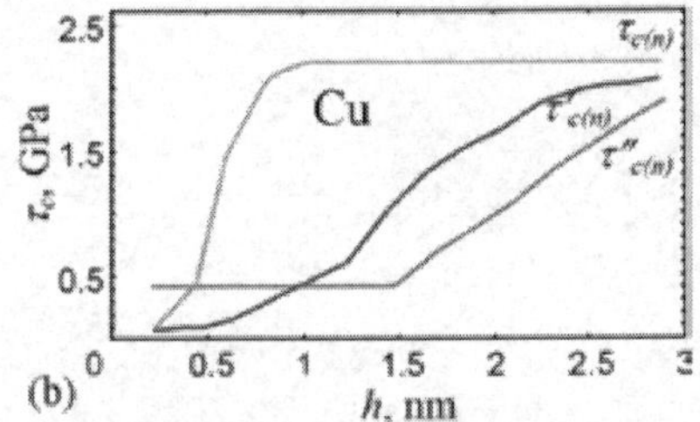

Figure 10. Dependences of the critical stresses $\tau_{c(n)}$ and $\tau'_{c(n)}$, for (a) nickel (Ni) and $\tau_{c(n)}$, $\tau'_{c(n)}$ and $\tau''_{c(n)}$, for (b) copper (Cu) on nanotwin thickness h.

According to results presented in **Figure 10**, the consequent emission of partial dislocations from locally distorted GBs is not favored. However, the critical stresses at which nanotwin generation mechanisms occur are highly sensitive to material parameters and the initial state of a locally distorted GB. Therefore, in the situations with other initial states and/or other materials, the consequent emission of partial dislocations from locally distorted GBs may be favored.

4. Conclusions

Thus, new specific micromechanisms of nanotwin generation at GBs in deformed nanocrystalline and ultrafine-grained materials were developed and analyzed. These micromechanisms describe the formation of deformation nanotwins at locally distorted GBs that contain segments being rich in GB dislocations produced by preceding plastic deformation. The micromechanisms of deformation twin formation occur through (1) the consequent emission of partial dislocations from GBs (**Figure 2**); (2) the cooperative emission of partial dislocations from GBs (**Figure 3**); and (3) the generation of multiplane nanoscale shear at GBs (**Figure 4**). It is found that the deformation twinning mechanisms (**Figures 2–4**) can operate in nanocrystalline and ultrafine-grained materials at rather high, but realistic levels of the stress (**Figures 6** and **8–10**). The suggested representations on generation of nanotwins at locally distorted GBs (see also [14, 15]) logically explain numerous experimental observations [11, 13] of generation of nanoscale twins at GBs in nanocrystalline and ultrafine-grained materials. These deformation twinning mechanisms illustrate complicated interactions between different deformation modes such as deformation twinning, GB sliding, and GB dislocation climb.

Acknowledgements

This work were supported by the Russian Fund of Basic Research (grant 16-32-60110) and Russian Ministry of Education and Science (task 16.3483.2017/PCh).

References

[1] Koch CC, Ovid'ko IA, Seal S, Veprek S. Structural Nanocrystalline Materials: Fundamentals and Applications. Cambridge: Cambridge University Press; 2007. 380 p. DOI: 10.1017/CBO9780511618840

[2] Hahn H, Mondal P, Padmanabhan KA. A model for the deformation of nanocrystalline materials. Philosophical Magazine B. 1997;**76**(4):559-571. DOI: 10.1080/01418639708241122

[3] Wang Y, Chen M, Zhou F, Ma E. High tensile ductility in a nanostructured materials. Nature. 2002;**419**(6910):912-915. DOI: 10.1038/nature01133

[4] Ke M, Milligan WW, Hackney SA, Carsley JE, Aifantis EC. Observation and measurement of grain rotation and plastic strain in nanostructured metal thin films. Nanostructured Materials. 1995;**5**(6):689-697. DOI: 10.1016/0965-9773(95)00281-I

[5] Feng H, Fang QH, Zhang LC, Liu YW. Special rotational deformation and grain size effect on fracture toughness of nanocrystalline materials. International Journal of Plasticity. 2013;**42**:50-64. DOI: 10.1016/j.ijplas.2012.09.015

[6] Gutkin MY, Ovid'ko IA, Skiba NV. Crossover from grain boundary sliding to rotational deformation in nanocrystalline materials. Acta Materialia. 2003;**51**:4059-4071. DOI: 10.1016/S1359-6454(03)00226-X

[7] Sansoz F, Dupont V. Quasicontinuum study of incipient plasticity under nanoscale contact in nanocrystalline aluminum. Acta Materialia. 2008;**56**:6013-6026. DOI: 10.1016/j.actamat.2008.08.014

[8] Bobylev SV, Morozov NF, Ovid'ko IA. Cooperative grain boundary sliding and nanograin nucleation process in nanocrystalline, ultrafine-grained, and polycrystalline solids. Physical Review B—Condensed Matter and Materials Physics. 2011;**84**(9):094103. DOI: 10.1103/PhysRevB.84.094103

[9] Ovid'ko IA, Skiba NV, Mukherjee AK. Nucleation of nanograins near cracks in nanocrystalline materials. Scripta Materialia. 2010;**62**:387-390. DOI: 10.1016/j.scriptamat.2009.11.035

[10] Chen M, Ma E, Hemker KJ, Sheng H, Wang Y, Cheng X. Deformation twinning in nanocrystalline aluminum. Science. 2003;**300**:1275-1277. DOI: 10.1126/science.1083727

[11] Wu X-L, Ma E. Dislocations and twins in nanocrystalline Ni after severe plastic deformation: The effects of grain size. Materials Science and Engineering A. 2008;**483-484**:84-86. DOI: 10.1016/j.msea.2006.07.173

[12] Zhu YT, Liao XZ, Wu X-L. Deformation twinning in nanocrystalline materials. Progress in Materials Science. 2012;**57**:1-62. DOI: 10.1016/j.pmatsci.2011.05.001

[13] Zhu YT, Wu X-L, Liao XZ, Narayan J, Mathaudhu SN, Kecskes LJ. Twinning partial multiplication at grain boundary in nanocrystalline fcc metals. Applied Physics Letters. 2009;**95**:031909. DOI: 10.1063/1.3187539

[14] Ovid'ko IA, Skiba NV. Generation of nanoscale deformation twins at locally distorted grain boundaries in nanomaterials. International Journal of Plasticity. 2014;**62**:50-71. DOI: 10.1016/j.ijplas.2014.06.005

[15] Ovid'ko IA, Skiba NV. Nanotwins induced by grain boundary deformation processes in nanomaterials. Scripta Materialia. 2014;**71**:33-36. DOI: 10.1016/j.scriptamat.2013.09.028

[16] MYu G, Ovid'Ko IA, Skiba NV. Generation of deformation twins in nanocrystalline metals: Theoretical model. Physical Review B—Condensed Matter and Materials Physics. 2006;**74**:172107. DOI: 10.1103/PhysRevB.74.172107

[17] Gutkin MY, Ovid'ko IA, Skiba NV. Crack-stimulated generation of deformation twins in nanocrystalline metals and ceramics. Philosophical Magazine. 2008;**88**:1137-1151. DOI: 10.1080/14786430802070813

[18] Lojkowski W, Fecht H. The structure of intercrystalline interfaces. Progress in Materials Science. 2000;**45**:339-568. DOI: 10.1016/S0079-6425(99)00008-0

[19] Ovid'ko IA. Nanoscale multiplane shear and twin deformation in nanowires and nanocrystalline solids. Applied Physics Letters. 2011;**99**:061907. DOI: 10.1063/1.3620934

[20] Boyer RD, Li J, Ogata S, Yip S. Analysis of shear deformations in Al and Cu: Empirical potentials versus density functional theory. Modelling and Simulation in Materials Science and Engineering. 2004;**12**:1017-1029. DOI: 10.1088/0965-0393/12/5/017

[21] Kibey S, Liu JB, Johnson DD, Sehitoglu H. Predicting twinning stress in fcc metals: Linking twin-energy pathways to twin nucleation. Acta Materialia. 2007;**55**:6843-6851. DOI: 10.1016/j.actamat.2007.08.042

Index